T0276107

# CLIMATE CHANGE:
## THE SHINY OBJECT IN THE ROOM

It's not what you think you know,
it's what you need to know!

RICHARD JONES

Archway Publishing books may be ordered through booksellers or by contacting:

Archway Publishing
1663 Liberty Drive
Bloomington, IN 47403
www.archwaypublishing.com
844-669-3957

ISBN: 978-1-6657-1556-0 (sc)
ISBN: 978-1-6657-1558-4 (hc)
ISBN: 978-1-6657-1557-7 (e)

Library of Congress Control Number: 2021923717

Print information available on the last page.

Archway Publishing rev. date: 01/13/2022

# Contents

# A Hint of Coming Attractions:

We are going to explore many issues in the coming pages, much of which is not necessarily common knowledge for the general public. To be able to get a good and firm handle on what we will talk about, you really need to put your Science Hat on. I recognize that Science is acquiring a bad name of late, but let's forge ahead anyway and tarnish that Shiny Object in the room, **Climate Change**. Topics include:

- Why the current focus on the reduction in the usage of our fossil fuel natural resource is a correct and worthy goal. However, even though it is the right thing to do, it is being marketed for the wrong reason. It's that pesky Shiny Object.
- Why there is no doubt that our weather patterns are changing, even as we speak. However, the jury is out, and in fact has not even been seated, when it comes to determining the cause of the changes. Remember, it's not "Climate," it's "Weather."
- The reasons for skepticism having to do with current climate models, and the modelers themselves. We will demonstrate how the vicarious nature of the chaos-driven random weather events really can never be predicted accurately and with any degree of validation. Making decisions on less-than-solid information ain't good.
- Some of the thermal aspects of our atmosphere and how those aspects are related much more to the Sun's interaction with the Earth's surface. It is much more about that huge, nuclear furnace that is just below our Earth's crust, than any "Green" issue.
- The absolute reliance that society, and the collective living organism that we are, have on Carbon and all its interrelationships with our lives and well-being. Essentially, we can't live without Carbon and everything it is in combination with. To attack it with abandon is reckless.
- The symbiotic relationship with our "bank" of fossil fuels, and what to do with what is left inside the Earth.
- The total consumption of those fossil fuels and what they are used for on a yearly basis, both globally and within the confines of the United States.
- Where those fuels are being used, what functions they perform, and the probability of being able to substitute the use of the abundant energy coming from the Sun.
- The need to understand that the symbiotic relationship with fossil fuels goes way beyond their use as a form of heat and transportation fuels.
- The issues that we must face when we move from a substantial use of fossil fuels to an almost total reliance on bio-energy and electricity generated from the Sun's energy or nuclear power.
- Finally, how to make sure that clear heads prevail in the coming years when we approach the problem of weaning ourselves from this symbiotic tie that we have with Coal, Natural Gas and Petroleum.

There is an educational, sales and marketing job that must be performed in the very near future in order to make sure our children, and our children's children have access to abundant electricity and fossil fuels where appropriate. Most importantly, we need to educate them on how to carry on rationally after we depart this Earth for good.

# Introduction

This is where I tell you what I'm going to tell you. Later, I'll tell you and then after that, follow-up with what I told you. Though in this current political climate "science" is getting some bad "press," the rational pursuit of truth through the scientific method is our staff of life. I am going to cause you to wade through a morass of concepts and data which will demonstrate the magnitude of the issues relating to the substitution of renewable energy of any sort for our vast, though finite, fossil fuel natural resource. There is really no way to "Dumb-Down" the explanation much more than I have done in the following prose. The facets to this problem are many and varied. To go much deeper, you would need a year's supply of No-Doze to make it through. Most importantly, if you come away with nothing else, you will realize that collectively "Climate Change," "Global Warming," and that pesky "$CO_2$" issue are nothing more than shiny objects being used by the **GND,** Green New Deal, movement for their own aggrandizement.

Weather, climate, the changing thereof and "what to do about it" have become so politicized as to hide in the fog all true, unbiased science and rational thinking. Climate Change, which really is the fluctuation of weather patterns as recorded historically, is not the real issue we need to address. It is the rapid depletion of our stored, fossil fuels and how to replace both their function as convenient sources of heat, and their use as raw material[1]. What I hope to accomplish with this document is the development of a foundation of good science and coherent thought upon which we can build a dialogue based on fact. Then maybe we can take advantage of our vast, engineering resources and the free enterprise system to solve the issue of depleting, fossil fuel resources. This depletion of a natural resource that we have no way of replacing is what we should be focusing upon, not our fluctuating weather patterns which is nothing more than the shiny object in the room. Let's eliminate emotion, political goals, financial gain, or wishful thinking from the equation. The overarching questions are:

- Who owns the fossil fuels that we extract from the Earth?
- Who should oversee controlling the use of those fossil fuels?
- Regardless of when the fossil fuels are consumed, they will still add to the $CO_2$ content of our atmosphere to a marginal degree and are not going away anytime soon [2][3]. When are we going to accept that fact and begin to address it and stop sweeping the inevitable under the rug? $CO_2$ is our friend, not our enemy.
- When are we going to study the positive effects of global warming (né: $CO_2$ buildup in the atmosphere)? I'll bet there are some, but politics are now in the way of the study.
- Who is winning and who is losing as a result of the **GND** narrative? Let's find out and expose it.
- Who started the $CO_2$ hue and cry and why?

Bottom line: let's make sure that, as time goes forward, we do the right thing for the right reason. This then allows capital to be expended prudently to arrive at solutions that have a reasonable and sustainable return on the investment. The term "Climate Change" is a shiny object that is a focal point distracting us from the real problem, and thus is counterproductive. As the saying goes, rightfully so, we really can't do much to change the weather.

# Greenhouse Effect

Before we go too deep into this treatise, we need to get the explanation of the **GE**: Greenhouse Effect out of the way. The **GE** has been around since time immemorial, or at least from the time that Earth acquired an atmosphere. From a very simplistic standpoint, the **GE** is nothing more than the process of our atmosphere acquiring (trapping) energy in the form of heat, regardless of its source, and the transfer of that trapped heat energy onto or close to the surface of the Earth. It is the temperature of the Earth's surface, and the atmosphere within a proximity to the Earth's surface, that drives our weather patterns.

The heat energy contained in and on the surface of the Earth is acquired in one of several ways; Solar Radiation, atmospheric re-radiation, atmospheric convection, or conduction from the Earth's Mantle into the Earth's Crust (surface). Essentially, the transfer of heat energy from the Mantle is the main stabilizing bias about which our daily temperatures fluctuate regardless of the season. Believe it or not, the other stabilizing element is our atmosphere through the **GE**. If these two factors were not present, probably we would not be able to sustain life on Earth.

Though not really a mystery, the **GE** is not easy to fully understand because it involves the blending of many scientific principles, the most noteworthy of which is Thermodynamics and the study of Psychrometry. Big words and esoteric terms, but let's dumb-it-down for those of us that have our feet on the ground.

## Side Bar #1, Love - 15

Quite honestly, the **GE** is all about radiation. All normal matter emits electromagnetic radiation, assuming that its temperature is above 0 °K or 0 °R (absolute zero). This radiation is a conversion process which changes a body's internal heat to electromagnetic energy. This is a constant Entropy process, and for the most part is reversable. More on Entropy later. Likewise, to some degree, all matter absorbs radiant energy, some compounds more than others[4]. It should be noted here that all matter absorbs and emits radiation at the same rate. In other words, matter's absorptivity and emissivity have the same value. Given this explanation, the exchange of radiant energy between our atmosphere and the surface of the Earth can be likened to a tennis player daily practicing his serve.

On the first day of his practice session, during the daytime, while he is at work, the Sun showers him with practice balls to fill his practice ball "bucket." At night, he uses these bucket balls to serve balls to a round disk down court from his position. If he can hit the disk, the ball bounces back to him so he can use it to refill his bucket. If he is in "equilibrium" with the number of balls he gets from the Sun each day, and the number of balls that he recovers from the disk-hits, then at the start of the next practice session, he has the same number of balls in the bucket.

However, if over time, the disk becomes larger in diameter, but his skill remains constant, more balls will be returned. He then will begin to accumulate balls to the point where his bucket overflows and he "serves" himself to an early grave.

Using this analogy, where the tennis balls are really heat energy, the **GND**ers cast the fear that our biosphere will accumulate energy to the point where humanity as we know it will cease to exist. However, I propose that there really is a "hole in the bucket" that the **GND**ers (and their complicit scientific community) are either totally unaware of, they know about but won't tell us, or they refuse to find because it doesn't fit their narrative. To obtain equilibrium, we need to find out the right balance between the two sources of radiation. Right now, we have no clue what those values are today, in 2021. That said, I expose more food-for-thought in **Parts B & C**

The ability of our atmosphere to acquire and radiate heat energy is inherent on all gases that compose our atmosphere. However, the majority of the individual, molecular impact, is contained in what are termed as Greenhouse Gases, GHG[5], basically due to their absorptivity/emissivity values. The composition of our atmosphere [6] is shown in *Table A-1*. Note that in the table, water vapor is not included in the "Dry Atmosphere" section. This is because our atmosphere is really a solute for water vapor and the psychrometric properties of this relationship are well known and well documented. The real issue is that at any moment in time, the amount of water vapor in the air is not constant, something over which we have little control, and literally varies from moment to moment.

A classic example of this is the difference in the "feeling" of the air as you exit an air-conditioned (cooled) space. The odds are extremely high that the air outside of the conditioned space contains substantially more water vapor on a percentage basis, than the air within the conditioned space. As a matter of fact, a primary function of an air conditioning system is to remove some of the moisture dissolved in the air, aside from changing the temperature. The big "however" here is that probably the composition of the other components listed above in the air both inside the conditioned space, and outside the conditioned space, remain the same.

Now, not all components of the Earth's atmosphere contribute to the **GE** equally. Though, as mentioned above, they all emit and re-radiate energy to one degree or another, but they all don't absorb and emit radiant energy at the same rate or amount. Only the GHGs contribute to the **GE** to any degree. In general, GHGs are not single elements, as is $O_2$ or $N_2$. They are multi-atomic compounds (molecules) such as Dihydrogen Oxide, $H_2O$ (water), Carbon Dioxide, $CO_2$, Nitrous Oxide, $N_2O$ and Methane, $CH_4$. *Table A-2* lists the most prevalent GHGs in the order of their contribution to the **GE**.

We really need to run a sensitivity analysis on the impact of each of the elements contained in *Table A-2*, however the data to do so is sketchy at best with the exception of moisture as it pertains to the air temperature. *Table A-3* shows how sensitive the contribution of water vapor dissolved in the air is to the average temperature of the surrounding air. This data is taken directly from the ASHRAE Psychrometric Chart No. 1, ©1992 ASHRAE. The weights shown are with the air at saturation conditions and atmospheric pressure of 29.921" Hg (14.696 psi).

As can be seen in *Table A-3*, the amount of moisture that the air can hold is substantially dependent upon the temperature of the air. A straight-line interpretation of that statement, given what we see in *Table A-2* above, is that more moisture means more reflective energy, which means more moisture, which means more heat. You can see where this could end up. One might also ask, with temperature fluctuations that occur over time, where does the moisture come from and where does it go when not in the air? This leads into the discussion of clouds; the most nebulous element used to predict future weather patterns.

# Clouds, as Viewed from 30,000 Feet

A cloud is defined as ". . an aerosol consisting of a visible mass of minute liquid droplets, frozen crystals, or other particles suspended in the atmosphere of a planetary body or similar space.[7]" Water vapor is the primary component of clouds, but other trace elements could be present. There are numerous categories (genera) of clouds; in fact, ten per convention. However, we are going to get up to 30,000 feet (no pun intended) and put all clouds into one of two arenas; those that provide an umbrella effect for the Sun's energy, and those that provide a blanket effect for the Earth's radiant energy.

Umbrella clouds reflect a bunch of the energy from the Sun but have little ability to absorb radiant energy from the Earth's surface and reflect it back to the surface. A proliferation of Umbrella clouds will tend to cool the Earth's surface and thus diminish any greenhouse effect.

Blanket clouds, on the other hand, allow most of, if not all of, the Sun's energy to reach the surface of the Earth and in addition absorb and reflect the surface radiant energy back toward the surface. A proliferation of blanket clouds will tend to increase the greenhouse effect.

Here's the problem with clouds, no matter of what sort. First, we have no real concrete historic data on the location, density or type of cloud. Second, we have no way of predicting exactly where they will occur next week, let alone next year or the year after. Finally, we have no way of knowing how clouds will truly affect the overall weather pattern given their obvious interaction with all the other changing weather patterns. For instance, if in the future, much of the cloud cover negated the absorption of surface radiation by $CO_2$, what would be the result?

## It's all about shades of grey

Basically, the consensus among "climate scientists" is that any variance in the prediction of our global temperature fluctuations is founded upon the lack of certainty surrounding cloud cover over the face of the Earth. Given the data presented in **Tables 1** through **3**, one can only say that a statement pertaining to the influence of clouds on any weather pattern is simply defining a shade of grey. We must take as gospel, for purposes of this document, the science surrounding the **GE** associated with the molecular gases in *Table A-2*. These gases will have a "blanket" effect on the radiant energy traveling to and from the Earth's surface. The basic question here is when does the accumulation of these "GHGs" get to the point where any fluctuation of cloud cover, no matter where it is positioned on the grey scale, can no longer mitigate the **GE** of those gases. The flip side of that question pertains to the probabilities of an unforeseen cloud cover feedback loop which forces substantial changes in weather patterns for very long periods of time resulting in localized living conditions that cannot be supported with known technology. Fundamentally, the **GND**ers are saying this is what $CO_2$ will do, so why not clouds also?

# Declarations, Definitions and Directions

Note that I will hesitate throughout this document to use the term "Climate Change" as a function of the Universe, because it really is a misused term, and most importantly, is not the issue. First, we probably can't do much to change our global weather patterns, so let's not waste time and valuable resources in the effort. Secondly, the real focus should be on the fact that we are now running out of fossil fuels at an accelerated rate. We can't make them inexpensively, and really shouldn't try. So, let's garner that precious resource, channel its use efficiently, and as a byproduct, figure out a way to maintain our glorious lifestyle.

At some point in the very near future, our use of fossil fuels will naturally begin to decrease from our current rate of consumption. The cost of extraction will by its very nature increase over time, driving the innovation of cost-effective substitutes (if the government stays out of the way). At some point in the (very) far distant future, usage of fossil fuels will asymptotically approach zero (0) and at that moment in time we will have much different things to bandy about. Quite likely, the issue of $CO_2$ accumulation in the atmosphere will have faded into oblivion a long time previously.

As you will see explained in the following treatise, the term "Climate Change" is simply a bell weather to which we need to pay some attention. Quite honestly, it's really the manifestation of the symptom, but not the problem. We are building up $CO_2$ in the atmosphere and it's a GHG. So be it. As you will see, that may be a good thing, or it may be a bad thing. We really have no clue what the result will be. To try to predict that "result" is pure conjecture, speculation and subject to overwhelming politization. Belief in the results produced by mathematical models is an article of faith. Remember GIGO. Garbage in, Garbage out.

## Side Bar #2, it's all about models.

The mentally vacant suits in our Nation's Capital currently are reacting to "investigative" propaganda developed by various individuals and groups that, in the whole or in part, state that the world is going to hell in a hand basket because of the buildup of GHGs. The propagators and believers of this rhetoric are the **GND**ers. The conclusions that are being touted as the truth were developed using mathematical models of one sort or another (probably). You can build all the mathematical models you want to predict the future, but they, and their results, are only valuable if, (a) the designer is non-partisan and has no agenda, (b) the designer is self- not politically- funded, (c) the designer is not self-aggrandizing, and most importantly (d) the designer uses input and design assumptions that make rational sense.

A good buddy of mine (and lots of other folks), Ronald Reagan, once said "Trust, but Verify." In terms of mathematical models, that phrase is modified to say "Trust, but Validate." When one tries to predict the future, the big problem is "Validation." Essentially, there is absolutely no way to validate any model that tries to predict the future. The permutations and combinations of events are infinite. Thus, any belief expressed, or action taken as a result of that belief, is totally based on faith and whichever direction the political winds blow.

There are various approaches to model construct, but I'm only going to focus on two; past trend line analysis, and finite difference models using nodal relaxation techniques. I've used and constructed both and they each have their place in the engineering world. However, both methodologies require a great deal of faith in the accuracy of historic data, and, as stated above, input assumptions that make sense. Trend line analysis modified with assumptions about the future are the easiest to build, but the least reliable. I liken this methodology to trying to write your name using a 10 (ten) foot long #2 pencil but only being able to hold onto the eraser end. The best you can achieve is some chicken scratching.

Nodal relaxation models on the other hand require substantial skill, a bunch of time - effort - money, and, in the case of trying to predict the change of weather patterns, a multidisciplined knowledge of most of the physical sciences. For example, if one were to try to predict the influence of the weather in New York City on the weather in Chicago, you not only have to know, at a minimum, about the obstructions to wind patterns, the thermal "chimney effects", the effect of the shifting of the tectonic plates below each city, the heat transfer coefficients of all of the elements of the Earth's crust between the two cities, etc. I could go on and continue to bore you, but I think you get the picture. If our current, very sophisticated weather predicting models employed by all the worlds weather services can't predict the exact path of a hurricane, how can we expect any model to predict what our weather is going to be in one year, let alone the year 2050?

## The Ultimate Model Test: Let's See Who Passes

Okay, here are the ground rules. The modeler can use any supercomputer in the world to perform the calculations. The beginning premises can only be the known weather conditions on 1 January 1920. The recorded composition of atmospheric $CO_2$ known at that time is the starting point, and the now recorded yearly increases in $CO_2$ from then up to the present are the only allowable input data. A perfectly constructed and 100% accurate model design will have results that unfailingly predict every hurricane and other natural disaster that has occurred from 1 January 1920 to the present time. Most importantly, it will accurately predict the temperatures of the Oceans, the atmosphere, the size of the Ice Caps, blah, blah, blah for every day of every year for each time-step of the model.

This is called "validation." If the model can't determine exactly what "happened," how can we trust it to predict what will occur? And OBTW, no cheating allowed.

## Some more Model questions

- What is the effect of population increases/decreases on weather patterns and the $CO_2$ content of the atmosphere?
- Have the models been exercised sufficiently to determine the ideal balance of radiation/ temperature/ etc.?
- What do the models show what happens when the global temperatures decrease for long periods of time? Given *Table A-3*, where will the water go? Oceans, lakes, glaciers, ice caps, . .?
- Again, referencing *Table A-3,* when the global temperatures increase, where does the water come from?
- Weather is like politics; it's local. When macro changes take place, like say in Australia, what is the trickle-down effect locally anywhere in the world? Are enough voters going to

be affected, positively or negatively, so that future elections are influenced one way or the other?

- How much "local" (micro) change needs to take place before the effect spreads and becomes "global" (macro)?
- What is the effect on the results of current models of the heat energy generated through the consumption of fossil and nuclear fuels? An estimate of the consumption of energy on a world-wide basis was in excess of 170,000 Terawatt-hours (TWh) in 2020 (see: "**Total Primary Energy Supply**"). That is a boatload of energy, in anybody's book, pushed into the atmosphere. Where is it?
- The most important question of all is what is the ideal balance of GHGs in the atmosphere, moisture content of the air, forestation, population positions and densities, fossil fuel usage, diminishment of existing solar flux through channeling it into energy carriers (see: *Table A-5*), etc., etc., etc.?

We do know that $CO_2$ is building up in the atmosphere, at least since the middle of the 18th century and into the early 19th century, primarily due to the industrialization of our society (see "**Greenhouse Effect**" above). We have been able to take advantage of the usage of the abundance of fossil fuels and the combustion thereof to blossom into what society is today. I think we can all accept that as a fact and therefore that is not what we need to focus upon. We need to first address and admit that fossil fuels soon will become a thing of the past. Secondly, we need to develop rational approaches to innovating technologies to replace the *function* of fossil fuels and the *products* that fossil fuels are used as raw material in the manufacturing thereof [8].

## Side Bar #3, along came C, O, and H.

In all fairness, we need to establish some facts of life that should be obvious to all of us here on Earth, but they have eluded some of those mentally-vacant folks that use the "News" media as their propaganda arm. There are three, basic compounds resident on earth that we need to pay some attention to: water ($H_2O$), oxygen ($O_2$) and carbon dioxide ($CO_2$). The carbon dioxide and water molecules, in the right combination within plants, and in the presence of sunlight (solar energy), produce glucose ($C_6H_{12}O_6$), the staff of life, plus oxygen ($O_2$) (Photosynthesis [9]). We then use the glucose and/or its derivatives, along with oxygen (respiration), to produce the energy that allows our plant and animal life to flourish.

This is the true cycle of life and continues day-in and day-out in a steady-state fashion. If we attempt to change this cycle in any way, through the man-made, artificial disturbance of this function, we have no way of determining the exact outcome. We are in that cycle right now and "nature" is adjusting nicely because any changes being made, by us humans or naturally, are happening over many years and in a very gradual fashion. If we attempt to hurry through some changes to this cycle, we have no clue what will happen, and it may be catastrophic.

During my research for this document, I happened upon a scientific paper titled "The Natural History of Oxygen"[10] by Malcom Dole. In this paper, he demonstrates some relationships that I had wondered about for a very long time; how did the fundamental elements that I talked about above come to be? The following shows what he and other scientists believe occurred during the (very) early stages of planet Earth when the only elemental building block available was hydrogen (H):

1. Hydrogen (H) molecules, through a nuclear reaction formed, Helium (He).
2. H and He, through another nuclear reaction formed Carbon (C).
3. C and He combined to form Oxygen (O).
4. O and H combined further to form Fluorine (F), which through a subsequent reaction formed Nitrogen (N).
5. This pretty much accounts for our present atmosphere, plus or minus a trace element or two.

As it turns out, the Oxygen thus formed (about 4.4 billion years ago), by its very nature, oxidized (combined with) everything in sight, including that new element Carbon, and it was not until photosynthesis came along later in the life of the planet, about 3.4 billion years ago, that the photosynthetic process was able to liberate Oxygen into the free, gaseous somewhat stable form of $O_2$ that we enjoy today in our atmosphere.

# Let's Define "Climate Change"

## Weather

". . the state of the atmosphere with respect to heat or cold, wetness or dryness, calm or storm, clearness or cloudiness"[11] (*noun, Merriam-Webster*) (In other words, what we all talk about, but can do nothing to change: *synopsis, Dick Jones*)

## Climate

". . the average course or condition of the weather at a place usually over a period of years as exhibited by temperature, wind velocity, and precipitation"[12] (*noun, Merriam-Webster*) (In other words, it's history: *synopsis, Dick Jones*)

So, let me sum this up. Climate is really an amalgam of lots of "weathers" over long periods of time at various locations on Earth. Therefore, it is data (read: fact), not conjecture, not speculation, not a wish, nor a means of getting a government grant to study something that's already happened and documented and nothing can be done to change or alter it (unless, of course, you need the money from the Government grant to take your next skiing trip to Aspen, or Snowmass, or wherever.) (*Summary statement, Dick Jones*)

For the last several years, I have watched and listened to mentally vacant "scientists" and "politicians" shove the "**Climate-Change**"/$CO_2$ hue and cry down the throats of our sorely uninformed world population; a self-imposed condition brought about due to the availability of "social media" platforms, government "study" grants, and a very gullible, attention hungry, print & TV media. This "group" has hidden under the guise of the **Green New Deal (GND)**. As an editorial note (which I can do because I am the editor-in-chief), I will not refer to the future change in weather patterns, regardless of geographic location, as "**Climate Change**." As explained above, climate is history and we can't change it because it's already happened. What we may, underscore *MAY*, be able to do is marginally change some, but certainly not all, weather patterns. The purpose of this document is to cast doubt upon "accepted" conclusions (sometimes disguised as the *conventional/current wisdom*), ask questions that should have been asked long ago, and make some logical suggestions about what we, as a society, should do going forward. Note that I hate bandying about "problems" without proposing probable solutions. I hope to accomplish this end using known facts, known science, reasonable assumptions, and just plain common sense. I will:

- Talk about our changing weather patterns (not "**climate**," because it's already happened; I repeat myself) and point out some obvious logic flaws currently being touted by the above mentally vacant, including the media mob,
- Talk about our carbon-based existence here on spaceship Earth and why it is necessary to understand the importance of $CO_2$ in our atmosphere,
- Demonstrate why any _short-term_ future push toward the use of "solar energy" is inherently distracting, expensive and counterproductive,
- Cast substantial doubt on the narrative pertaining to the (**GND**ers, Government-funded) use of electric vehicles as an immediate, hell-for-leather, solution to the mitigation of climate change through the reduction in $CO_2$ emissions from the use of fossil fuels.

# Our Changing Weather Patterns (né: Climate Change)

Unfortunately, the use of the words "**Climate Change**" have become the label for a cause, heralded by a bunch of "Crusader Rabbits," rather than a descriptor for a previous, current or most probable future series of weather patterns. Don't get me wrong. Our climate is changing and has changed; hour-by-hour, day-by-day, year-by-year. By the above definition, it must change because it is an agglomeration of historic data driven by a series of chaotic events over which we have had no control. I can see it every day when I access Weather.com to see if I need an umbrella that day. Not only do we have a diurnal cycle to deal with, but variances within the diurnal cycle, and variances within the variances. The diurnal cycle exists due to the 23.4° tilt of planet Earth and its rotation about the Sun. Glacial periods (ice ages) and interglacial periods (warming cycles) are caused, believe it or not, by slight variances in the geometry of the orbit of the Earth around the sun; apparently every 100,000 years or so. That is one helluva flywheel to deal with. Conventional wisdom speaks and I have no reason to discount this theory.

All these phenomena are occurrences that again, we have no control over. It is the micro-variances that take place within the yearly cycle that drive the current narrative because that is what we experience in our daily lives. It's that tornado that touched down one county over that we focus on in order to define "**Climate Change**." We have no (micro-based) clue what really occurred 250 years ago because we weren't there to experience the weather, and the scientific instruments that would have been necessary to chart the climate were less than state of the art as compared to today. Just in case the **GND**ers of the world are listening, we had no satellite imagery to rely upon when the Declaration of Independence was being signed (independent of the 1619 Project). The scientist within me says loud and clear that the flywheel effects of the thermal masses (more on that later) associated with planet Earth (Ocean currents, atmospheric streams, Tectonic Plate shifts, etc.) are so great as to dwarf any short-term (year-to-year, decade-to-decade) variances in cloud cover, the rotation of the Earth and its effect on wind patterns, atmospheric composition, the topography variances across the Earth's surface, or the vicarious nature of the butterfly flight patterns in San Luis Obispo, CA.

## Side Bar #4, houses on stilts?

We are led to believe that one of the issues that those of us who live on or close to the coast of one of our oceans has to face is what happens if all the ice in the world melts? I'm talking ice caps, glaciers, even the ice cubes in your soda. I'm not even addressing any increase in the level of the ocean because I think that will be the least of our worries. It's all about the Earth's atmospheric and surface temperatures. If the surface of the Earth gets hot enough to melt all the ice in the Arctic and the Antarctic, what do you suppose the temperature will be in your own back yard, let alone the Equator or along the coast of Florida?

The real issue that needs to be addressed is the determination of why weather patterns change on a micro level at all. Are we currently in a micro cycle, or within a macro cycle? Do we really have the tools and the data base necessary to even start the prediction process? What effect does the "Greenhouse" we live within really have?  What we have done so far to address this issue is, to me, a dismal failure.  As stated above, we can't even determine 24 hours in advance the exact path of a hurricane, the exact location of the next tornado touchdown, or the exact amount of snowfall in Broken-down Chimney, Montana.  Consequently, how can any rational person expect an accurate predication of weather patterns, and the results of same, one hundred, fifty, or even two years in the future?  To even consider that any predictions that far in advance are anything but voodoo science, witchcraft, or at best, calculated guesswork, is self-deluding.  Most importantly, is it prudent or even rational to make substantial, and possibly catastrophic, financial and societal decisions based on less than concrete information?  Finally, no matter how hard we try, can we really, in our fondest dreams, do anything about the "change" or cause the "change" to change?

# Carbon and its Importance

My guess is that the founders of the **GND** and the rest of the crowd that worships at that altar, including our less-than-mentally-nimble president (sleepy Joe), has/have absolutely no clue that we are, and have been since the beginning of life on Earth, a carbon-based society. Probably, if one asked any **GND**er to name three things within their sight that contained carbon, they would stutter their way through maybe 2. Planet Earth cannot continue to exist, as we know it today, without carbon in all its forms. It is contained in the concrete you walk on, the steel in your car's engine, your golf club shaft and head, and your fingernail. Carbon is the 15th most abundant element on Earth and is the second only to oxygen as the most abundant element in the human body[13]. Because of Carbon's ease in forming polymeric structures and organic compounds within Earth's temperature ranges, it serves as the common, elemental tie to all known life on Earth.

We are led to believe, through the narrative of the **GND**ers, and I might add, a complicit media, that the villain within the plethora of carbon compounds is $CO_2$. It is classified, right or wrong along with such compounds as methane ("natural gas"), nitrous oxide (a sort of byproduct of combustion), ozone (lots of sources, combustion included) and water vapor as a GHG, the function of which is to trap heat energy within the atmospheric envelop surrounding Earth (see "**Greenhouse Effect**" above).

## Side Bar #5, setting a trap.

We need to talk about trapping heat. If our atmosphere did not trap heat, and oh-by-the-way absorb heat, especially during the day, we would have a climate exactly like the surface of the Moon (you know, Luna). If the Sun is shining on the surface of the Moon, it's hotter than a tin roof in Arizona at high noon on June 21 of any year. If the Sun is not shining on the Moon's surface, it's close to the temperature of Outer Space (darn near absolute Zero {~4°K}). This is because the Moon has no appreciable atmosphere.

You can demonstrate this trapping phenomena to yourself. During daytime, regardless of the weather condition (assuming it's not raining) water does not form (condense) on the surface of anything. However, at night (given the ideal atmospheric conditions) with even just a few clouds in the sky, "dew" (or "frost" depending upon the time of year and your latitude) forms on all surfaces that "see" the sky dome. If the surface is hidden from the sky dome, due to substantial overcast of clouds, being under a carport or shaded by a tree, no moisture forms. The scientific reason behind this phenomenon is that the sky dome at night tends toward a temperature approaching absolute zero, buffered only by our atmospheric envelop which during the day was marginally heated by the Sun but mostly by the heat energy flowing convectively or radiantly from the Earth's surface.

Now, at the onset of the night (read: The Sun goes down), the minimal amount of heat energy stored in the atmosphere (see: **Parable Hour** below) quickly begins to radiate into outer space. The saving grace which keeps us from literally freezing to death (as we would on the Moon) is the fact that the energy stored on the surface of the Earth begins to radiate toward and convect into the now rapidly cooling upper atmosphere and in the process, heats the atmosphere, though marginally. Some of this heat is absorbed by the upper atmosphere and re-reflected to the surface, but not enough to keep the surface temperatures of the unshaded areas from cooling rapidly.

When the temperature of any exposed surface cools enough to reach and then drop below the dew-point temperature of the surrounding air mass, the moisture dissolved in the air condenses onto those cold surfaces and, Voila! Dew. In essence, at dusk, the sky dome looks like an object (a really big object) that is about 25 +/- degrees below the temperature of the Earth's surface and continues to drop in apparent temperature throughout the night. Thus, any surface that "sees" the sky dome will radiate and convect heat energy to the sky dome up to the point at which that surface reaches the "dew point" temperature of the surrounding air.

The point here is that the atmosphere is a buffer protecting the stored energy on and within the surface of the Earth, day and night. If the atmosphere did not pick up the stored heat on the surface, and/or reflect heat energy back onto the surface of the Earth, then when the Sun goes down, we would reach environmental temperatures that could not sustain life. Therefore, some re-reflection/re-radiation is good. The real question is how much re-anything is bad?

**A Worthy Note:** If real science were to be performed about $CO_2$ and its role in the greenhouse effect, then we should study the timing of Dew Point Formation. A simple explanation of this experiment would be to gather some local data on the atmospheric psychometric conditions, the surface temperatures of the surrounding Earth/structures, correlate these with the $CO_2$ content of the atmosphere, and determine how long it takes to form Dew on the exposed surfaces. In other words, if $CO_2$ had a real, short-term, measurable and catastrophic effect on the trapping of heat, then the increase in $CO_2$ in the atmosphere should slow the formation of Dew. Time to write that grant application and send it ASAP to the **GND** Society, Washington, D.C.

Per "**Greenhouse Effect**" above, water dissolved in the atmosphere has by far the greatest and almost overwhelming "Greenhouse Effect." However, to be fair, $CO_2$ has a greater influence on the trapping and re-radiation of heat than all the _other_ atmospheric compounds combined [4]. The **GND** narrative then goes on to state that this trapping of heat, regardless of the cause, is forcing detrimental changes in our weather patterns which, in the (real) extreme, will eventually wipe humanity off of the face of the Earth. What they really are saying is that there has been an increase in the trapping of heat, which is then reflected back onto the surface of the Earth, which in turn forces (maybe to a measurable degree) a change in our weather patterns. There are many factors that influence the surface temperatures on Earth, reflected/radiated energy being only one of those factors, albeit very minimal.

Okay, let's do a little scientific investigation of the history of $CO_2$ and its functionality. First and foremost, plant growth could not exist without $CO_2$. It, along with water, sunlight and some basic minerals thrown into the mix creates, through "carbon fixation," the carbohydrate-based structure for trees, grasses and broccoli. In fact, since the carbon cycle began billions of years ago, carbon has been "banked" into the components which formed the fossil fuels that we are employing today to carry on our daily lives.

It took millions of years to extract the $CO_2$ and water ($H_2O$) from the atmosphere and the surface of the Earth and accumulate it into plant and animal life. Then, with an additional application of time, lots of subsequent heat and pressure were applied to finally alter and condense those life forms, once dead (don't worry, PETA, nothing live was used), into a form of kinetic, chemical energy that we could use at a later time and place. The consumption of that dormant (though finite), stored energy source, which is giving us a quality of life not seen before in history, is putting $CO_2$ back into the atmosphere at a rate greater than it is being redeposited into minerals, the oceans, and plant and animal life.

Water, as in the lakes and oceans, is a solute for $CO_2$ (read: "carbonated beverages"). The oceans form a huge bank for $CO_2$. As a matter of fact, the $CO_2$ concentration in the oceans is in the range of 45 to 50 times that of the atmosphere[14]. This massive concentration is a double-sided coin because the solution of $CO_2$ in the ocean tends to increase the acidity which disturbs some of the natural, biological cycle of the ocean biosphere[15]. That is the subject of a future treatise that is not covered here, and probably has little effect on changing weather patterns, but certainly is worthy of an informed dialogue.

On the other hand, the historic concentration of $CO_2$ in the atmosphere is worthy of investigation as part of this treatise. As of the end of 2020, the overall average concentration of $CO_2$ in the atmosphere was approximately 415 parts per million, or 0.0415% by volume. The concentration in 1960 was a little less than 320 parts per million, or less than 0.032%. In general, it is thought that due to human activity, almost all of which is due to the combustion of fossil fuels, the concentration of $CO_2$ in the atmosphere has increased since the onset of the industrial age; around the turn of the 18th century[16].

In 1750, it is proposed that the $CO_2$ concentration was 280 ppmv (0.028% by volume) as compared to May of 2021 when the average concentration was 419 ppmv (0.0419%), according to the National Oceanic and Atmospheric Administration.[17] This represents a 50% increase in concentration when comparing the increase to the 1750 levels. Interestingly enough, the same article states that the concentration in May, 2021, has not been seen for about 4 million years when (in theory) the Earth's temperature was 7 °F warmer and the sea levels were 78 feet higher than today. Me thinks that just the concentration of $CO_2$ in the atmosphere may not be the only factor influencing our climate, past and present.

## Side Bar #6, how's your utility bill these days?

Assuming that the above percentage figures are correct, when compared to 1960, our average increase in $CO_2$ concentration is 30% (0.0095/0.0320). Has the temperature in your backyard increased 30% (on the average)? How's about your heating bills? Your cooling bills? Maybe you haven't noticed the change at all. Shame on you! However, probably we could get a pork-funded grant to study that issue. Dial #250 and ask for "The Squad."

Regardless, we need to put these data in perspective. 500 million years ago, the concentration of $CO_2$ in the atmosphere has been projected to be upwards of 20 times what it is today. This concentration apparently decreased to about 4.5 times the current concentration during the Jurassic period and then decreased precipitously about 49 million years ago to levels approaching our current 0.03% to 0.04%[18]. Given the laws of nature, that matter itself cannot be created nor destroyed, and assuming that the amount of carbon that was created as part of the formation of the Earth hasn't changed, then a bunch of that carbon that was available 500 million years ago ended up in sea shells, limestone, mineral deposits and fossil fuels. It certainly isn't in the atmosphere in the form of $CO_2$.

Okay, this then leads to the conclusion of this portion of the treatise, and the inevitable questions that arise.

- How many tons of carbon are still in the ground locked up in the form of fossil fuels? What percentage of what we already have consumed is that total weight?
- Without a change in any of our usage patterns today, when will the price of fossil fuels increase to the point where the free enterprise system develops cost effective alternatives? How much $CO_2$ will be created between now and then and where will it end up?
- When those fossil fuels that have yet to come to the surface are consumed to extinction, and converted into $CO_2$, how much $CO_2$ is created and where will it go?
- Given that approximately 50% of the $CO_2$ created through the combustion of fossil fuel remains in the atmosphere, what is the effect on the local weather patterns going to be? Where does the other 50% go[19]?
- If the $CO_2$ in the atmosphere, especially within the boundary layer close to the Earth's surface, increases to the point where it begins to accelerate the growth of all green-growies, what is the effect on the weather?
- If I put 100 lbs of $CO_2$ into the atmosphere, through the combustion of fossil fuels, theoretically, 50 lbs remain and 50 lbs are absorbed (somewhere, somehow). Can we extend this analogy to say that if I put 2 lbs of $CO_2$ into the atmosphere, will 1 lb. remain there? If that is the case, then we can never reach carbon neutrality unless, (a) we create no $CO_2$ in anyway, and/or (b) we artificially remove the excess $CO_2$ from the atmosphere.
- If less $CO_2$ is better, then let's start removing it from the atmosphere right now and see what happens.
- What has the increase in human and animal population throughout the world done to the balance of $CO_2$ in the atmosphere? How much of the increase in $CO_2$ is attributable to this factor? How much can this effect increase before we annihilate ourselves?
- Most importantly, since we have no data on what our life would be like with, say, twice to three times the atmospheric concentration of $CO_2$ that we are experiencing today, who's to say it would be better or worse? To put this in perspective, how high can the ocean rise before it melts all the ice and completes its volume increase via thermal expansion? How much land mass is lost, and do we adapt to being a "water world" before the next, inevitable ice age? Can we accept the fact that this cyclic Ice Age phenomena is unavoidable no matter what we do?
- There is a feedback loop involving the consumption of fossil fuels used to maintain the climate control of our living condition which should be considered. If our climate becomes so temperate and stable, and we develop a lifestyle such that neither heating or cooling will be required for comfort conditioning, what happens to the rate of consumption of fossil fuels?
- If you have other questions, you can use this space to fill them in.

# Total Primary Energy Supply (TPES)

Okay, enough about the weather/climate per se. We now need to focus our dialogue on our global energy supply and sources of heat. World-wide TPES is a sum of all energy consumed in the world in one year. Primary Energy is a form of energy that we as humans have not modified or converted in any way to make it more "usable." We really need to define, in a lay fashion, two terms.

## Enthalpy

This thermodynamic term is used to describe the total amount of heat-associated energy contained in a body, or a collated group of bodies, considering the volume and pressure of those bodies. This term does not address latent chemical energy which we will deal with below using different terminology. Enthalpy is expressed in terms of Btu (British Thermal Unit) or Wh (Watt-hour), generally as it relates to the mass of the bodies in question. A simple example would be a cup full of water sitting on a kitchen counter at room temperature. By its very being, it has a given amount of energy (enthalpy) contained within. Logically, we know this because if we wanted to cool that body of water, to the point of, say, freezing it, we would have to remove some, but not all, of the heat energy it contained.

## Entropy

Now this thermodynamic term is a little more difficult to understand, so bear with me. Fundamentally, this term applies to our existence. Its premise is that within an environmentally closed system, if work is done on one part of the system by another part of the system, the entropy of the system remains the same. For example, let's pretend that a system was composed of you, as a person, and a hermetically sealed device containing a cylinder full of air with an enclosed piston. The system, at rest, is in equilibrium with no outside influences, and therefore each component has a measurable enthalpy, and the system has a measurable amount of entropy.

Now, you use the piston to compress the air in the cylinder, the result being that the air changed both in volume, and in temperature. The volume of the air decreased and the temperature of the air increased, just like in a diesel engine. In other words, work was done, by you, on the air resulting in a transfer of energy which manifested itself in a temperature increase and stored kinetic energy vested now in the air in the cylinder. The enthalpy of the air was increased, while your total enthalpy decreased by a like amount. However, the entropy of the system did not change because nothing was added to the system, and nothing left the system. This is what is called a "constant entropy" transaction. However, this transaction is not reversible, as some are, because part of the enthalpy that you lost was chemical in nature and nothing within the system can be used to replace that chemical energy. Therein lies the crux of our problem as a society. Fossil fuels are the chemical energy that can't be replaced.

Keep with me on this. Even though I can generate a dynamite example of our Federal Government's failed taxation policy using the process described above, I'm not going there. I'm more interested in studying the "system" within which we all live. To be blunt, we are losing "entropy," and as a result we are heading toward a cliff, but not the **GND**-described cliff.

Our TPES is for the most part finite, except for the Sun, and in some circles, nuclear fuel. Remember, the Sun drives everything on Earth; wind, solar electromagnetic energy, biomass, hydro, etc. Therefore, at some point the Sun, and the remaining nuclear fuel is all we will have left. *Tables A-4 & A-5* show the universe of Primary Energy Sources and the conversion processes involved to get them to be "Energy Carriers." It is the Energy Carriers that we use to conduct commerce and our daily lives.

# Energy That Drives Our Lives

First and foremost, we have only three available sources of energy that can and do drive our daily lives. Our spaceship Earth cannot exist without taking advantage of one or more of these sources. Our greatest, and most "renewable" (albeit finite) is the Sun. The second is nuclear, which also is finite and of (reasonably) known quantity. The third is fossil fuels; finite and pretty well documented.

We are going to talk about how our non-renewable sources of energy are being used, with a focus on the generation of electricity and its projected use as a universal automotive propellant. In reality, except for the use of natural gas, petroleum or coal as a raw material, almost all uses of our non-renewable resources can be replaced, or at a minimum, augmented by electricity. For instance, coal and petroleum are transformed into "coke" which is a prime component in the smelting of iron and the manufacturing of steel. Plastics have as a feed stock, petroleum and natural gas products.

Alright let's get some facts/data out of the way. Various fuels and methodologies are used in the generation of electricity, as documented in the world-wide data from 2018 presented by the International Energy Agency.[20]

- Non-Renewable energy sources, 74% (Coal – 38%, Natural Gas – 23%, Nuclear – 10%, Oil – 3%)
- Renewable "Energy" sources, 25% (Hydro – 16%, Wind – 5%, Solar – 2%, Bio-fuels – 2%)
- Other (undefined), 1%

When it comes to the generation of electricity using non-renewable energy sources, they must employ a methodology which relies almost exclusively on the generation of steam through the application of heat in order to drive a steam turbine, which in turn powers an electrical generator. The exceptions being the use of natural gas or fuel oil to power engines of sorts which, in turn, drive electrical generators. This is not a 100% efficient process. Far from it. Because the amount of electric usage world-wide is so huge, the dimensioning employed is in terms of **terawatts** and/or **terawatt-hours**, or TWh. 1 TWh is the equivalent of 1 billion kilowatt-hours, or enough to power over 60,000 average, Florida homes for one year.

For the year of 2008, total Primary Energy consumed world-wide at all power plants for the generation of electricity was 51,158 TWh, which was 36% of the total for primary energy sources (TPES) of 2008. Electricity output (gross) was 20,185 TWh, at an efficiency of 39%, and the balance of 61% was generated as heat. A small part, 1,688 TWh, or about 3% of the total heat input, was utilized at co-generation heat and power plants. The in-house consumption of electricity and power transmission losses were 3,369 TWh. The amount supplied to the final consumer was 16,809 TWh which was 33% of the total energy consumed at the power plants. [21] Net electrical production jumped to 26,700 TWh total from all sources in 2018. Therefore, assuming some degree of linearity, 74%, or 19,760 TWh was produced by steam generation in power plants for end-use consumption.

# The Usefulness of Nuclear Fuel

I'm going to address the usefulness of each of the sources mentioned above, focusing on nuclear first. When we think of nuclear fuel, the first element that comes to mind is Uranium. "Uranium's average concentration in the Earth's crust is" . (depending on the reference) . 2 to 4 parts per million, or about 40 times as abundant as silver" [22]. An interesting reference, I must say. Nuclear energy/fuel comes in multiple forms, the pursuit of the definition of those forms and their derivation is beyond the scope of this treatise.

Nuclear energy, per se, is heat (used as described above) which is a result of the decay of fissionable material. We have yet, outside of the hydrogen bomb, figured out a way of emulating the Sun, which of course is a huge fusion furnace. Therefore, taking advantage of current technology, it is fission we must rely upon, which only serves to be able to generate heat, which in turn is used to generate electricity using Nuclear Reactors. "The development and deployment of these Nuclear Reactors continue on a global base. There is increasing interest in these power plants as they are powerful sources of $CO_2$-free energy. In 2019, 440 nuclear power reactors produced 2,586 TWh (billion kWh) of $CO_2$-free electricity worldwide, more than the global installations of solar and wind power combined" [23].

Even then, we still have to "throw away" 60+/-% of the heat generated through fission onto the Earth's surface or into the atmosphere (it's inherent in the process), while only getting the remainder in the useful form of electricity. To be fair, **_ALL_** the energy used to generate electricity, no matter the fuel, eventually ends up heating something which eventually heats the atmosphere. Seems wasteful, but that's the result of that ugly thing called physics (you know, "science"). To close this one out, we can't make plastics, change the oil in our cars, or grow crops with only nuclear fuel. Therefore, its usefulness is finite and quite limited.

# The Usefulness of Fossil Fuels

The next most useful source of energy is the abundance (currently) of fossil fuels. The three forms of fossil fuels have been used by some pretty smart people to give us the lifestyle currently enjoyed by us U.S.-ians today. Granted, it took many millions of years, and lots of the energy generated by the Sun over those millions of years, to "bank" the latent energy in those fuels. Currently, all we are doing is withdrawing that non-renewable energy at a rate that is so substantial that at some point in the (very near) future we will run out and have to rely upon other sources and/or technologies in order to survive as an entity. I have no clue when that time will come, but it will come. Hopefully, the free enterprise system, and our capitalistic society, will gradually come up with cost effective, and market-driven solutions that will allow humans to continue to exist. Time will tell, but I ain't going to be around to see it happen.

More on the use of our fossil fuel resource later; in detail, I might add.

# The Sun; The Last but Not the Least

Okay, this leads us to the final source, the Sun. The amount of energy reaching ground-level, on Earth through a cloudless sky, day in and day out, is about 1 kW per square meter perpendicular to the Sun's rays. In other words, in one hour, the Sun provides a total of 1 kWh of heat and other electromagnetic energy per sq meter, assuming that same square meter remains absolutely perpendicular to the Sun's rays for that entire hour, and no clouds get in the way. Obviously, this is not a 24/7 function. In general, the calculations used going forward in this document factor in various degrading factors to arrive at a "usable" value for the average amount of heat/energy that we can count on from the sun on a yearly basis.

## SIDE BAR #7, that pesky Btu:

We need some down-to-Earth frames of reference. The above 1 kWh/m², expressed in English parlance is 3,412 btu/m² or 317 btu/ft²/hour. Further reference: it takes 1 btu to raise one (1) pint (pound) of water one (1) °F. More later during "**Parable Hour.**"

Keep in mind that no "square meter" is stationary with respect to the Sun's rays, and only receives that level of energy once per year. This is called "direct normal radiation" and is basically in the form of electromagnetic energy of many wave lengths, visible and otherwise, and must be converted in order to be useful in other forms, like electricity. Obviously, it's pretty difficult to transmit the Sun's rays throughout the transmission and distribution system that currently we use to power our light bulbs, so the Sun's energy must be converted into an Energy Carrier (see: *Table A-5*).

To continue with the Sun's contribution, we must realize the other two functions that the Sun performs, both of which we can't do without. The first is weather. Without the sun, we would have no weather, or even a change in weather, because the Sun is a continuous, and highly reliable, source of heat. It produces the heat that forms the wind, clouds, rain, snow, blah, blah, blah. Without wind, we would have no diversity of climate, nor wind generated pollination of plants. Without the Sun, we would have no plants, which of course would result in no food. Okay, I'm stating what should be obvious, but probably isn't to those folks that I mentioned at the beginning of this treatise.

To convert the sun's energy into something useful in the form of heat or electricity requires substantial capital investment and is inherently inefficient. The sun's energy can be used in one of three ways to directly, or indirectly, produce electricity; wind generation, hydroelectric generation, or photovoltaic generation. Using the Sun's energy in an attempt to try to generate electricity the form of heat only, is a fool's errand.

One of the key descriptors used to analyze the usefulness of a power-generating system is its capacity factor. Basically, this is a comparison of the 24/7 "name plate" capacity when compared to its usable output over a finite period, such as a month or a year. All the afore mentioned methodologies have substantially lower capacity factors than base-load fossil fuel plants or nuclear power plants, which in general can have capacity factors greater than 90% to 95%. Theoretically, this category of plants could have capacity factors of 100%, but the diurnal usage of electric power within the distribution system, and the need for periodic maintenance and refueling force plant output reductions and downtimes.

- Wind Power. The sun provides the energy to generate the geographically oriented, atmospheric high- and low-pressure regions which result in wind. This is an intermittent source of power which only can be relied upon at certain times of the day or year, and only in certain geographic locations. Historically, the data show that wind turbines have capacity factors of between 15% and 50%[24], again depending upon the constraints. The average annual capacity factor for wind generators in the U.S. between 2010 and 2015 was 30% to 34%. The cost effectiveness is another issue altogether and must take into account site location, maintenance needs and such things as inclement weather conditions similar to that experienced in Texas during the winter of 2020.
- Hydroelectric Generation. Though it should be obvious, hydroelectric generation is as susceptible to climate variations as all other solar-driven technologies. The water behind the dams is replenished through rain fall and water shed runoff, neither of which can be predicted with any accuracy.
  - ○ Dominion Power in VA came up with a not necessarily unique, but very practical, solution to promote consistency of supply. In Bath County, VA, they constructed two reservoirs, one higher than the other, separated by a hydroelectric generating/pumping station. They employ off-peak generating capability (primarily at night) to pump water from the lower elevation "pond" to the higher elevation "pond." Then during the peak times of the day, water can flow from the upper reservoir, through hydroelectric generators and into the lower reservoir. There are obvious efficiency losses (12% to 15%), but the reliability is known and the process can reasonably be used when needed.
- Regardless, due to climate variations, the capacity factors for known hydroelectric facilities range from 45% best case, to 23% which is the probable case for most facilities.
- Solar Power. Finally, we have the case of photovoltaic cells which are aggregated into large "farms" in order to gain some economies of scale. They only work during daylight hours and the conversion efficiencies are dismal when compared to the total energy contained in the direct, normal solar radiation. In addition, due to such factors as cloud cover, dust and general climate variances, documented capacity factors range from 29% best case to the more probable case of 12% to 15%. In essence, a lot of real estate must be allocated in order to obtain a reasonable amount of power for distribution into the grid.

We have a lot of valuable (taxpayer funded) resources; money, time, manpower, chasing the problem of how to take advantage of the Sun's abundant energy. A corollary to that is how we mitigate the effects on our weather attributable to the Sun and reflected energy from those pesky, GHGs. This leads us to "Parable Hour."

## Parable Hour

About noon on a bright sunny day in our Nation's Capital, a rather influential, self-aggrandizing member of our House of Representatives, was at her previously vandalized desk when her lackey du jour burst into her presence and exclaimed "We have found the solution to the problem!" With (contrived) tears of joy, she launched herself from her previously vandalized office chair and exclaimed "Show me, show me, do!" Her now guide and problem solver, né lackey, rushed her out to the Capital parking lot and showed her two identical, boxes about the size of the old phone booths previously prevalent on every street corner and outside every bar.

"Ma'am, on the top of each of these boxes is a one-foot square window which will let Sun-light in, but not out, and the boxes are sealed and constructed of 100%, insulating panels so that the heat from the Sun cannot escape. We filled one box with air, and the other with water just a few minutes ago. Now we have to wait for about an hour for the results."

"Great," she said, "because I have to get my hair done and at the same time, draft some meaningless legislation that I can shove down the throats of all those idiots on the other side of the aisle. Come and get me when you are ready, I'll be the only one in the salon"

An hour later, the lackey drags her back to the demonstration and exclaims, "See, air is the answer!" Somewhat mystified, she queried, "Why?"

"See the thermometers on the side of each box? The one on the box that only contains air reads 370°F, while the one filled with water hasn't moved from the 80°F where it started. We need to call a press conference and announce this to the world!"

As they rush back to her office, she elbows the lackey and asks under her breath, "tell me again. What was the problem?"

The above parable is a demonstration of how the **GNDer's** approach to "Climate Change" is all about shiny objects and not about real science. If read carefully, the Sun's rays would have been almost perpendicular to the glass window on the top of the boxes. The otherwise, unimpeded energy from the Sun would have entered those boxes and, in the hour that it took to get her hair done, the mass within each box would have absorbed the 317 btu mentioned previously in **Side Bar #7**. Once that energy entered the boxes, it had nowhere else to go, so it dissipated into heat and applied it to the only mass available.

## Thermal Density.

Thermal density plays a major role in the discussion of weather and climate. Thermal Density is defined (for this document at a minimum) as the amount of energy needed to be applied to a compound in order to raise one (1) cubic foot of that compound, 1°F. The Thermal Density of water is 62.3, steel 53.1, concrete 22.5, and air 0.018. The thermal density of water is by far and away the best compound to be used to store thermal energy without substantial increases in temperature. Air has a thermal density that is 0.029% of that of water. In other words, it is a dismal failure when it comes to storing and mitigating energy fluctuations when temperature is the ruling metric.

As will be seen later in the verbiage, understanding the thermal density of the Earth is of great importance. The thermal density of the outer Earth's crust varies from the oceans at ~62, to the solid crust forming our land masses at approximately 30 to 35. Therefore, per °F, there is a boatload more energy stored in what is on the surface of the Earth as opposed to the air surrounding our bodies.

The above example demonstrates the uselessness of using air temperatures to try to define long term weather patterns. The typical phone booth used has a volume of about 60 ft³. When filled, it will contain 3,744 lbs of water or 4.52 lbs of air. Given the amount of energy input, in that hour, the air increased in temperature 291°F, while the water only increased 0.085°F. And you wonder why the air in your car gets so hot so quickly. Bottom line here is that to assume that air temperature has anything, in and of itself, to do with our changing weather is a fool's errand. Look somewhere else for the solutions.

Probably, given that the word "Climate" is implicit in the title of this document, we should spend some time on the driving forces of our weather patterns, which by definition and through the documentation of historic data define what we commonly call our climate. Our climate is driven by the temperature of, and the energy contained in, the surface of the Earth. Heat travels in only one direction; hot to cold. That is a physical law and no matter how it is spun, it is still the law. Heat travels from the (very) hot surface of the Sun, to the (relatively) very cold surface of the Earth. Heat also flows from the ~11,500°F core of the Earth, 3,900 miles below the surface, through the Mantle layers and the crust to the surface of the Earth.

The Earth's Mantle[25], which extends from 1800 miles below the surface, at a temperature of ~4,900°F, to 250 miles below the crust, where it reaches a temperature of ~2,000 °F. The Mantle is heated both by energy in the core and fissionable, nuclear reactions within the Mantle itself. This heat buildup forces massive, convective currents of force which adjust the tectonic plates, resulting in changes in the surface temperature of the Earth. The amount of stored heat energy in the Earth is truly unconscionable and extremely difficult to put in perspective. This massive amount of heat energy changes "position" constantly, and on a random basis erupts onto the surface and affects, sometime dramatically, our Earth's surface temperatures.

In closing, we need to keep in mind that as the penetration of any solar-driven technology (primarily, if not exclusively focused on electricity generation) increases beyond a certain limit, the extent of which is beyond the scope of this document, a means of storage must be employed. The usage demand of nuclear plants must be close to 100% of nameplate capacity, which is why they are used primarily as base-load satisfaction facilities. The usage demand and supply demand of fossil fueled plants can be matched quite nicely, thus their flexibility to be used during both peak and off-peak time periods. This is not the case with the above solar driven technologies. A common method of storage is batteries, but hydroelectric generation techniques such as the Bath County example can be employed in order to reduce cost and take advantage of our abundant higher elevation regions.

# U. S. Total Primary Energy Supply

**"Total Primary Energy Supply** (TPES) is the sum of production and imports subtracting exports and storage changes[27]." Essentially, it is the summation of all the energy that we use, here in the United States, on a periodic basis, regardless of the source of the energy. We talked previously about TPES, but only on a world-wide basis. The World's consumption of energy is 162,494 TWh (*Table A- 4*). The U.S. consumption is 27,240 (16.8% of world supply) (*Table A-6*). The following tables and figures demonstrate the configuration of energy usage in the U. S. only. Summation of usage will be presented in terms of Quadrillion Btu (Qbtu) and Terawatt-hours (TWh).

| | |
|---|---|
| 1 Quadrillion (Qbtu) | $= 10^{15}$ Btu |
| 1 TeraWatthour, TWh | = 1 billion kilowatt-hours (kWhs) |
| | = 293.08 Qbtu |

The graphic method of presentation used in the following figures is the use of circles whose total area represents their designated portion of a total. This method should not be thought of as linear when comparing circle diameters. Area is a second-order function of the diameter - radius relationship. For instance, a circle with a diameter Y has an area X. To represent a circle with an area X/2, Y becomes 0.71Y, not Y/2. 0.71 Y is the square root of 0.5 Y.

# Explanations of Tables A-6 – A-8

Overall, the data presented in the U. S. Energy Information Administration (EIA) Monthly Energy Report for April, 2021 that is used for all tables, is presented in terms of its source, the energy content of the contribution of that source, and the ultimate destination (Retail Consumption) for the source energy. In order to provide consistency in dimensioning of the energy, regardless of source or destination, I have adopted the usage presented by the U. S. EIA. The data will be presented as either Quadrillion btu (Qbtu) or Terawatt-hours (TWh), or both on the same table.

## Table A-6: U. S. Energy Consumption by Sector, 2020

*Table A-6* presents the three basic sources for all the energy consumed in the United States and is subdivided by Sector (see *Figure A-1*). The Fossil Fuel and Nuclear sectors should be obvious. However, the Renewable Energy Sectors require some elaboration.

- **Hydroelectric** This term is used to describe the harnessing of energy from flowing water. The water flow can be vertical as in Hydroelectric Dams or Tidal basin flow, or horizontal as in Wave or Water Wheel movement.
- **Geothermal** This includes Hot Water or Steam extracted from Geothermal reservoirs below the Earth's surface, or the use of Geothermal Heat Pumps which utilize the near-ambient temperatures on or near the surface of the Earth.
- **Solar** This assumes the conversion of Direct Solar radiation into either heat or electricity utilizing various conversion techniques.
- **Wind** This is the use of the Kinetic Energy contained in the weather-induced movement of air to drive mechanical systems such as pumps or electric generators.
- **Biomass** This addresses the incineration of waste products other than Fossil Fuels. Included are all wood-based products (sawmill waste, tree waste, etc.), municipal solid & liquid waste, industrial and agricultural waste, agricultural products grown specifically for incineration, and biofuels such as ethanol, biodiesel, etc.

## Table A-7: U. S. Energy Consumption for Electricity Generation, 2020

Approximately 38.5% of all the energy consumed in the United States is devoted to the generation of electricity (see *Figure A-2*). *Table A-7* breaks down the contribution of each of the sectors shown in *Table A-6*. The "% Elec" shows the percent contribution of a particular fuel/source to the total contribution of all fuels/sources. "% Total" takes the same sector consumption and compares it to the U. S. total energy consumption. Note that some of the fuel/sources are used for both heat and power generation, and those uses are delineated in other tables and figures as CHP, Combined Heat and Power.

## Table A-8: U. S. Electrical Energy Distribution by Sector, 2020

Approximately 35% of all the energy consumed in the generation of electricity is sold at "retail" (see *Figure A-2*). In general, the retail sales include those "sales" that result from usage within the generating facilities, and the transmission and distribution losses associated with getting the power to the "meter." The resulting 65% accrues to the surface of the Earth and/or the atmosphere in the form of heat. Please note also that the vast, over 99%, of the Electricity sold at retail also ends up in the form of heat deposits. The total retail sales of electricity account for 13.5% of our consumption of energy, while the thermal losses amount to 25% of our consumption.

# Fossil Fuel Usage Reductions

This is a very sticky wicket because what I am presenting as "Probable Case" and "Ideal/Possible Case" Energy Consumption Reductions are pure guesswork on my part. I will justify why I chose a percentage figure for the reduction shown, but that "guess" is still a guess, and as the saying goes, "it's anybody's guess." We know that a reduction has to take place. The issue is how much, when, and what replaces the function of that which we are removing from contention. Below is the summary of "how much" and "when." The "replacement" issue will be attacked after that segment is concluded. *Figure A-3* delineates consumption of Petroleum, *Figure A-4* is devoted to the distribution of Natural Gas, and *Figure A-5* shows how the usage of our Coal resources are allocated. The following tables and their explanations reference each figure in question as noted.

## Table A-9: U. S. Petroleum Consumption by End Use, 2020

*Table A-9* presents the 13 basic categories used by the EIA to segregate Petroleum Energy Usage in the United States. The assumed percentage of reduction for each category is shown on the line following the descriptor. The "Most Probable Reduction" is my assumption as to what can rationally be accomplished within the next 10+/- years. The "Ideal Case Possible Reduction" is my assumption of what can be rationally accomplished within the next 30+/- years. The totals shown on the last line are the **resultant** percentages of **current** usage, obviously from a different viewpoint than all other line items. For instance, in the Most Probable Case, we will have reduced our usage by 20.80 %, and in the Ideal Case, we will have reduced our usage by 60.84 %. Each of the line items in this table are the collation of the assumed percent reduction in each category of each sector within the category. A summation of the impact on each of the 5 sectors is shown in *Table A-10*.

## Table A-10: U. S. Petroleum Consumption by Sector, 2020

*Table A-10* presents the data for the 5 basic categories used by the EIA to segregate Petroleum and all other Energy Usage in the United States. The assumed percentage of reduction for each category is shown on the line following the Sector descriptor. Most of the reduction takes place in the Transportation sector, as can well be imagined. The remainder of the reductions assume that the petroleum product reduced was used primarily as a source of heat or mechanically, as in gasoline or diesel engine fuel. No petrochemical feedstocks have been assumed to be affected. The totals shown on the last line are the **resultant** percentages of **current** usage, obviously from a different view point than all other line items.

**Note**: There exists a 4% difference in heat content data for Petroleum Usage in the U. S. when comparing Tables 1.3 and 3.6, EIA Monthly Energy Report, April 2021. The reason for this difference is not readily apparent in the literature. A possible explanation could focus on the ascribing of heat value in the various sectors of Petroleum usage, given the different qualities of raw petroleum consumed in each sub-sector. Rounding errors of 1% or less cannot account for all of this discrepancy. Essentially, the data presented in this document are all "in the ball park" and should be viewed from 30,000 feet as opposed to ground level. The distribution of Petroleum usage is the largest culprit of this discrepancy. The remaining sectors are relatively solid and correlate nicely within the EIA Documentation.

## Table A-11: U. S. Natural Gas Consumption by Sector, 2020

*Table A-11* presents the data for the 8 basic categories used by the EIA to segregate Natural Gas Energy Usage in the United States. The assumed percentage of reduction for each category is shown on the line following the Sector descriptor. Most of the reduction takes place in the Electric Power, Residential and Commercial sectors, as can well be imagined.

Lease & Plant Fuel and Pipeline & Distribution are affected little because whatever natural gas needs to be supplied must be produced and transported. More than likely, this is a very generous estimate of what will take place in the future.

Not all Industrial CHP can be replaced by electricity, simply due to imbedded industrial plant infrastructure. It is assumed that all non-CHP[1] is used for product feedstock which cannot be replaced with current technology.

In addition, I have assumed that all Residential and Commercial usage is for heating purposes and therefore can be replaced with Electric power. The only catch to that approach is the enhancement of the Electrical transmission and distribution system; an issue that will be addressed later in the document. Electric power usage of natural gas will not go to zero at almost any point in the future. Again, this issue will be handled later.

The totals shown on the last line are the **resultant** percentages of **current** usage, obviously from a different viewpoint than all other line items.

## Table A-12: U. S. Coal Consumption by Sector, 2020

*Table A-12* presents the data for the 3 Sectors and 5 Sub-sectors used by the EIA to segregate Coal Consumption in the United States. The two main areas of reduction are the CHP sector and the Electric Power Sector. The CHP assumptions used are as above. Electric Power Generation is a different subject, which will be expanded upon later in this document. Suffice to say, Coal Power Plants are considered base load plants and can only be replaced by other base load plants, or substantial "fly-wheel" technology such as batteries or hydroelectric storage techniques. Coke cannot be replaced because it is a critical element in the manufacturing of steel and other materials that use carbon as a structural component.

The totals shown on the last line and other lines with **bold print**, are the **resultant** percentages of **current** usage, obviously from a different viewpoint than all other line items.

# Summary of Fossil Fuel Consumption Reductions

We need to understand the overall impact of the reductions that I have postulated in *Tables A-9* through *A-12*, and *Figures A-3* to *A-5*. As stated above, the reductions may or may not be possible. However, to gain insight on what might happen under those assumptions, we need to approach the problem rationally.

Our current Residential, Commercial and Industrial infrastructure is based solely on the use of either fossil fuels as they are presented to the retail buyer, or electricity as it is used to accomplish numerous tasks; heating, lighting, generation of mechanical power, etc. Some fossil fuel usage cannot be replaced using electricity in substitution. A classic example is the use of Coke in the production of steel. The most obvious areas that electricity can be substituted for fossil fuels is to generate heat at the retail level, and as a form of energy in transportation.

However, both areas of substitution come at a price. We must substitute some form of energy derived directly from the Sun and convert it into electric KWhs so that it can be transmitted to the fixed-in-place nodes of retail usage. In addition, since in general the geographic location of this auxiliary generation is not in close proximity to either the existing Electrical grid structure or the end user, we have to create additional transmission and distribution facilities, which of course impose a loss penalty on the generated product.

- We must be careful about the pricing of the kWh at the retail, or end-of-use level. A kWh, once it enters the Transmission and Distribution network is fungible and cannot be distinguished as to its source. Therefore, to price a kWh that is used to replace an incremental amount of fossil fuel consumption on the assumption that it was generated using renewable sources, is a form of discrimination that probably will be tested in the Courts. Therefore, all conversions from fossil fuel usage to electricity must stand on their own merits and be implemented solely on economic terms. Any Government incentives to make the switch are prone to failure and will be ripe with corruption.
- This is not to say that the cost of each kWh is the same. I have spent the better part of my professional career designing, analyzing and implementing the use of Time-of-Use (TOU) and "demand-based" electricity rate structures, especially as they relate to the residential consumer. There is a substantial advantage for the Electric Utility to be able to sell a kWh "off-peak" as opposed to "on-peak," and to be able to control the diversified demand on the grid during times of need. We will go into those economics later in this document.

Aside from the generation of electricity using fossil fuels, the largest consumer of fossil fuels is the transportation sector. Electricity generation consumes 5,974 TWh in fossil fuel energy, while the transportation sector consumes 6,789 TWh. In theory, heavy on the "theory," we could eliminate all the transportation usage, but in a practical sense, there really is no cost-effective way to eliminate a large percentage of fossil fuel usage in the electricity generation sector. More on that aspect later when we talk about TOU rate structures and Grid supply flexibility.

## Table A-13: Summary of the Reduction in Fossil Fuel Usage

*Table A-13* summarizes the three areas of fossil fuel usage that have the capability of being replaced by electricity. Obviously, this replacement is incremental to all current retail usage of electricity; a subject that we will dwell upon later. Regardless, the replacement of the "Electricity Generation" and "Heat" aspects are relatively straightforward other than the fact that a great deal of investment at the end, or retail, usage must take place in order to implement the substitution. The replacement in these two areas must be done as a result of economics as opposed to government intervention through subsidies or other discriminatory practices which of course are subject to fraud and misuse.

Transportation, however, is a completely different matter. Electricity cannot be directly inserted into the existing supply chain nor the existing end use infrastructure. If electricity is to be substituted as a fuel for transportation purposes, it will affect every citizen of the United States. Trillions of miles per year are driven using current, fossil fuel technology, in millions of automotive forms. Eventually, in order to implement this conversion 100%, each current Fossil Fueled vehicle must be replaced by an electric vehicle of one type or another. The infrastructure impact alone is enormous. We have in place a supply network for our transportation fossil fuels that over time will be replaced with a like supply network for the electric "fuel." As will be seen below, this is not a trivial matter.

Currently, transportation fossil fuels fall into two categories: Petroleum and Natural Gas. Reference *Table A-10* and *Table A-11* above, the total usage of fossil fuels for transportation purposes is 6,806.9 TWh per year, with Natural Gas being 0.26% of the total. Thus, any reduction in Natural Gas usage, though measurable, is not that significant in the overall scheme. It will reduce $CO_2$ emissions, and contribute less to pollution, but it is the usage of petroleum products that will make the greatest impact, both from a usage standpoint and the disruption of our current supply chain infrastructure.

A great deal of effort has been devoted to the development of Electric Vehicles (EV), most notably, passenger cars. A substantial portion of the development has been subsidized by the Federal government, both from the standpoint of direct grants to automobile manufacturers, and tax "breaks" and other incentives provided to the ultimate consumer for the purchase of the EVs. For EVs of any variety to be universally accepted, the driver must be market economics, not government subsidies. So far, the EV industry has enjoyed a lot of cherry-picking, but that tree is starting to run out of fruit.

Before we go further, we need to address the impact of changing from fossil fuels to electricity. *Table A-14* below demonstrates the probable increase in electricity generation in order to accomplish the massive change in our usage patterns.

## Table A-14: The Reduction in Petroleum Usage, Requires Some Explanation

The EIA has given us total usage of petroleum for the purpose of transporting anything and everything. What we really don't have an absolute grasp on is the actual number of miles driven by any class of vehicles. We have figures that are truly estimates because there is no reporting mechanism in place which reads the odometers at the beginning and ending of each year on each vehicle. Let's hope that process is never implemented.

Given the lack of certainty, I developed the analysis assuming a bandwidth of usage ranging from an annual Miles per Gallon (MPG) of 25 to a low value of 10. Again, for purposes of demonstration only, I assumed all fuel was gasoline and had the same conversion rate of 33 kWh per gallon, per the note[a]. Using the example above and tracking through the table focusing only on the bold outlined cells, we can come to some conclusions as to an order of magnitude pertaining to the additional TWhs per year that need to be generated to supplant the previous usage of fossil fuel.

As stated previously, I have assumed a most probable reduction within 10+/- years of 1,878.81 TWh in fossil fuels used for transportation purposes. Using the stated conversion factor, this translates into 1.42 trillion miles driven, assuming an efficiency of 25 MPG with fossil fuels. I studied many different EV manufacturer's literature (BMW, Volkswagen, NIO, BYD, Tesla, Ford) to determine the EPA sanctioned published efficiencies for their EVs. The data is published as Miles per kWh, and the average among all vehicles studied was 3.63. To give you a frame of reference, an equivalent fossil fuel gasoline consumption is 0.76 miles per kWh assuming an automobile efficiency of 25 MPG.

Taking this analysis further, we would have to supply 392 TWhs at the "pump" in order to supplant the fossil fuel. However, there are losses getting that electrical energy to the "pump." Those losses accrue to the transmission and distribution of the electricity from the source, regardless of the source. The source power must be converted into transmittable power and then transmitted, at a quantitative loss, to the final destination. Within our current grid and source structure, these losses are accepted to be 7% to 8% of the generated power, or 7.5%. In other words, currently, we must generate 1.08 kWhs in order to deliver 1 kWh "at the meter." If our sources in the future are not at the current generating facilities, then that generated power also suffers transmission and distribution losses, at least equivalent to the current losses. The result of losses in both "grid" systems then means that we must generate 1.17 kWh at the alternate source in order to deliver 1 kWh to the meter. Therefore, the 392 above now becomes 458 as shown in the table.

# The Conversion to Electricity and its Impact

The Obama-Biden administration tried their hand at coming up with a subsidized solution to the **GND** by funding Solyndra, a failed (and extremely expensive, taxpayer-funded) experiment intended to develop cost-effective solar collectors, basically for commercial, rooftop installation.

The $CO_2$ generation portion of the "Green" narrative is the driver for the "need" to force electric vehicles, of all makes and models, down our throats. The current desire of the **GND**ers is to be rid of fossil fuels between 2035 and 2050, depending on who is speaking and to what audience. I'm coming at this from 30,000 feet and the generalities that I will use are in fact generalities. However, the percentages quoted can be easily verified through extensive and comprehensive study. The data is all available, it is simply a matter of focus and collation.

Okay, let's get back to the theme; electric cars/vehicles and other substitutions. The current administration is championing a cause that is doomed to failure, both from an implementation standpoint, and economically. Keep in mind, I'm painting with a broad brush and the absolute value of the assumptions used could be questioned, but the thrust of the theme is well founded. Much research needs to take place in order to apply defensible numbers on what I am going to put forth, but that research is doable, albeit incredibly time consuming.

Leveraging from the information presented in *Table A-13* and *Table A-14,* and the subsequent discussion, within 25 to 30 years, we are going to have to generate a possible 1,270 TWhs in addition to our current generating capability just to service the needs of the EV industry, in the worst case. Referring to *Table A-15*, we can expand this analogy to **all** fossil fuel substitutions. The best case in that time frame is an additional 5,809 and the worst case is 6,571 TWhs per year. Our current generation capability is 3,664, which also is the capacity capability of our current transmission and distribution infrastructure. In other words, in ~30 years, we would have to increase our sources and the grid structure by 159% to 179%. Obviously, this is not a trivial task.

The technology readily available to us currently that we can use to supplant fossil fuel usage is either the use of Solar Collector Farms, or Windmill Farms, both of which have substantial costs associated with their implementation. Not only is the hardware expensive, but the real estate is also costly. The use of the term "Farm" should not be taken lightly, as you will observe.

*Tables A-15, A-16 & A-17* demonstrate the task ahead of us if we are to replace our usage of fossil fuels in a sane and cost-effective manner. Continuing our tracking of usage reduction from *Table A-15*, we take the same worst-case new generation requirement shown as 6,571 TWhs and move to *Table A-16*. We see that not only have we increased our current generating capacity by an additional 179% of what is in place now, we have also started to chew up some mind-boggling real estate.

The absolute best case for real estate acquisition is if solar panels could be placed side-by-side with absolutely no gap in between. The most probable instantaneous efficiency of the current Solar Panel production is about 19.5%, per the manufacturers published literature[28]. This translates into an ideal 1-hour output of 0.184 TWh per mile² per year, assuming 100% density of the Solar array/farm. Continuing with the ideal-case analogy, assuming 10 hours per day of sunshine, and a capacity factor of 40%, the array will generate 0.737 TWhs per mile² per year. When the capacity factor dips to 12% (the likely minimum observed), then production drops to 0.221 TWhs per mile² per year. I used these two constants to build *Table A-16* and expanded the narrative to assume panel densities of 50% and 25%

The use of wind turbines is approached from a slightly different angle. The turbines don't rely on sunshine per se to generate power, so their site location is what rules more so than solar panel installations. As noted above, there is about 0.3 MW per hectare (ha) available for wind generation[29]. A ha is a square about 328 feet on a side. At a 25% capacity factor, wind turbines require about 5.88-mile² per TWh produced per year. Density only comes to play with wind because for the most part, the turbine itself consumes little real estate, and what is under the turbine blades, if low enough, can be used for other purposes. The real limiting factor is the transmission and distribution infrastructure and its impact on the cost effectiveness of the site chosen for the Farm. The assumption, of course, is that the site was chosen judiciously with respect to its historic wind direction and speeds.

*Table A-17* gives us a frame of reference in order to allow us to understand the massive land area that will be required to produce the incremental power we need to produce in order to supplant the usage of fossil fuels. Viewing the results presented in *Table A-16* you can see that with the probable case reduction in usage, a 12% capacity factor operating on real estate with a 25% density, the area needed is 36,184-mile². This is equivalent to the entire land area for the State of Indiana. Wind Farms, at a 25% capacity factor would require more land than the state of Maryland. I've driven in and across both of those states and I must tell you, that's a lot of land area.

# Let's Connect the dots

All right, it is now time to connect all the dots that I have created and come to some conclusions, vis a vis, possible solutions. We here in the United States currently face political, economic, social and just plain structural problems. The solution(s) to these problems is/are difficult to achieve without some real belt-tightening. The current Biden administration is either, (a) totally unaware of our problems, (b) incapable of understanding the problems, or (c) ignoring the problems in the hope that they will disappear on their own. Make no mistake, the biggest problem, our depleting natural resources, does exist and if we continue to ignore it, it will be at our own peril. Enough chit, chat. Let's connect some dots.

I don't know who originally coined the phrase, but the truth is that an army travels on its stomach. Without food, a soldier has no energy to fight. The United States travels on its energy resources. All else aside, but not unmindful of all other "important" issues, we as a nation cannot survive one day without having energy, in one form or another, readily available for use. Right now, we have no real long-term strategy when it comes to our use of energy, nor a stewardship plan for our energy resources, regardless of their nature. What I hope to accomplish as we go forward describing the "Dots" and their connections, is to clarify the problem in layman's terms, and develop the fertile ground necessary for the planting of thoughtful "seeds" that hopefully will grow to the solutions that we need going forward.

## Dot #1: Our Weather Patterns

Do we really need to do the research pertaining to our changing weather patterns, under a non-partisan umbrella and completely out of control of whichever political party happens to be in power at any moment in time? As I have stated above, we can't predict the exact occurrence of any form of weather, be it the path of a hurricane, the exact touch-down point of a tornado, or the amount of new and accumulated snow fall in Aspen on 25 December 2021. Therefore, we really don't have a "micro" handle on what causes weather fluctuations.

We're pretty good when it comes to the "macro" level, starting with "this winter, it will be cold in New York and Hot in Sydney." I'm being facetious, but you get the drift. To get much beyond this level, like accurately predicting the path of a hurricane, an incredible amount of historic data must be gathered, correlated, and then applied to almost micro data in real time. This is a daunting task, and quite honestly beyond the ability of us humans.

The bigger question before the house is whether we really need to know more about predicting weather patterns than we already have in place. It almost becomes a "so-what" issue. Right now, we know enough to button up for a possible landfall of a hurricane, and that the atmosphere is ripe for tornado touchdowns. That said, how can we rely upon weather models that look years into the future, with no validation whatsoever as to their accuracy, when it comes to developing wholesale social, economic and structural decisions?

I suggest we stop this gross, long-term prediction nonsense and focus our resources on issues that really will go far to change our lives for the better now and into the far future. As the old saying goes, politics is local and so is the weather. Couple that with the fact that we can't change the weather, then you have a recipe for moving this issue not only to the back burner, but off the stove top altogether.

## Dot #2: $CO_2$ and other GHGs

I keep going over in my mind a point that I brought up earlier in the tome. We really have no earthly clue what the weather was like in 1776, when in theory the $CO_2$ content of the atmosphere was quite a bit less than the current, 2020 value of 415 ppmv, or 0.0415% of all the atmospheric components. Probably, it was in the range of 200 to 300 ppmv, quite a bit less than current concentration. At another time and place many millions of years ago, per the literature, it was 20 times what it is today, or 8,300 ppmv (0.83%). What do you suppose the weather was like then?

I've told you above that $CO_2$, $O_2$, $H_2O$, Sunlight and a handful of assorted minerals are glued together to form our staff of life – green growies. All that fossil fuel in the Earth's crust is a result of a lot of those elements coming together, flourishing, dying, rotting, changing shape and form. Was this done for our benefit, or are we just taking advantage of a happenstance? Given the good that came out of lots (more than we have today) of $CO_2$ in the atmosphere, and probably lots of warm weather and sunshine, why are we so paranoid about an incremental rise in the content of any **GE** gas in the atmosphere? Who is to say that more $CO_2$ won't enhance our lives as opposed to leading us to an early apocalypse and global doom?

We need to bury the $CO_2$ buildup rhetoric as it pertains to the consumption of fossil fuels. Rather, we need to accept the fact that with the proper stewardship of our fossil fuel natural resources, our usage will diminish over time and this shiny object called "Global Warming" or "Climate Change" will fade from view. We have bigger fish to fry, so let's focus on things we can do something about, and that truly enhance our way of life.

## Dot #3: "Our" Petroleum Resources

Who owns the fossil fuels under the geographic boundaries of our Nation? Is this reserve a National Treasure, or is it anybody's to take as they might? Our National Parks are National Treasures and are controlled and protected mightily from desecration and other forms of vandalism which over time would destroy their beauty and usefulness. Imagine what would happen to Mount Rushmore if every tourist was able to take a chunk of Abe Lincoln or George Washington home with them.

Once a modicum of fossil fuel comes out of the ground and enters the supply chain, it is a fungible commodity and we have no idea who "owned" it to begin with. I am not going to dwell extensively on who owns what and where, because my guess is that it is a rat's nest of legal and political dealings, past and present, that would take another tome the size of this one to sort out. What I want to accomplish is the understanding that the resource is finite, the life-blood of our society and very existence, and currently is being used and consumed capriciously, without thought toward the future, and is a political hammer endlessly searching for a nail.

We must set aside political infighting and self-aggrandizing moves on the part of our "ruling" elite and act like adults. Every once in a while, I see an adult pop his/her head up within the maelstrom of the DC swamp, but they soon fade from view because the drive-by media has a different, self-serving agenda, and fails to give credit where credit is due. I have some thoughts on this issue of ownership, but it is integral with other "dots" that I must get out of the way first. Thus, you will have to wait to see the "film-clip" at the end.

## Dot #4: Fossil Fuel Priorities

Assuming you were paying attention, you saw in the various breakdowns of fossil fuel usage that each fuel; petroleum, natural gas and coal, have primary, secondary, tertiary and beyond uses. Some of those uses can only be accomplished with a particular fuel while other uses are transparent as to the fossil fuel consumed, such as electricity generation.

**Natural Gas** – One of the products created with $CH_4$ that are difficult to synthesize otherwise is its function as a feedstock in the manufacturing of carbon-based polymers such as polyethylene and nylon. I have classified this within the data as "Industrial Non-CHP." Other than that category, it is devoted to providing direct heat or the production of electricity. The one area that will be difficult to deal with is its use as the primary source of energy for space heating. The imbedded distribution infrastructure is one thing to deal with as I will do below. However, the 800-pound gorilla is the political and financial problems surrounding the imposition of a change from fossil fuels to electricity, especially within the residential sector. This is not a trivial matter and to sell the need for the change involves a massive marketing campaign targeted to almost every citizen of our great Nation.

**Coal** – This one is pretty easy. Most of our coal usage is the fueling of our base-load electric generating power plants and other CHP uses. Only 8.06% of our coal consumption goes toward Non-CHP and Coking production. However, supplanting the use of coal to generate electricity to feed the current grid structure also is not a trivial matter. As will be seen below, coal along with nuclear power is the primary fuel for our baseload generating capability. Inquiring minds wonder why the **smart** folk in China and India, who are privy to the same "information" from the Intergovernmental Panel on Climate Change, IPCC, continue to fuel their thriving economies with Coal. This is just one of the essential pieces in the "What Do We Do When the Well Runs Dry" puzzle that we are tackling right now.

**Petroleum** – This is the "Biggy." You can see in *Tables A-9* and *A-10* that the uses of petroleum in all forms spans many sectors and uses within those sectors. Its usage as a means of generating electrical energy is trivial; just 0.55% of the total consumption. Transportation, however, is the major user with 68.9% of the total consumption. Most of the Residential and Commercial usage, 5.1% of the total consumption, is used for the generation of heat and can be supplanted with electricity if necessary. However, most of the industrial usage, 25.5% of the total consumption, cannot be replaced with an alternative currently on the horizon. These uses encompass asphalt construction, lubricants, coking, etc.

## Dot #5: Energy Carrier Infrastructure

Currently, we have three energy carrier infrastructures which serve our energy needs quite nicely, thank you. The petroleum sector has pipeline and surface transportation means that will be used less in the future. Pipelines, truck and rail transportation move both raw crude oil and refined products to their eventual use at the refinery or the "pump."

Natural gas and other light distillates such as propane and propylene also move through an extensive pipeline, rail and truck network. The ends of this network for the most part is comprised of "meters," above ground and sub-surface storage facilities.

The transmission and distribution facilities for electricity are all around us. Overhead transmission lines, power pole transformers and gigantic voltage reduction substations are an everyday part of our lives. Long distance transmission of electricity using overhead facilities has in the past caused a hue and cry vis a vis electromagnetic causality of brain vacancy. There must be a lot of overhead transmission infrastructure in D.C. Just sayin'.

All funnin' aside, all this infrastructure came at a cost. The capital associated with the construction of that infrastructure is being amortized into the price that we pay at the "pump" and the "meter." Don't forget the shareholders that also need a return on their investment. Natural Gas and other light distillates are for the most part under the regulation of state corporate commissions, but they are also subject to overall supply and demand, and therefore we see periodic, though minor, fluctuations in the retail price. Electricity, on the other hand, because of the presence of long-term supply contracts, and known fuel mixes, tends to remain relatively stable, retail pricewise.

If we reduce the flow of petroleum and gas products via the existing infrastructure, then the unit cost will tend to increase due to the capitalization component of the retail price. Electricity, on the other hand, is a horse of a different color. The existing transmission / distribution system is limited, at times, in its capability to handle the existing usage, let alone piling any more on top of what is there.

We have before us a classic example of one of the problems that we will face in the future when the use of fossil fuels to heat homes needs to be replaced with electricity, especially in our northern climes. Most of these residences, and the feeder circuitry to get to the meter, do not have the metering nor service entrance capacity to handle an additional 25 to 40 kW of power. This may mean that a 125 Amp service entrance may have to double in size to a 250 Amp service. As you can imagine, this, too, is not a trivial undertaking because the "ripple effect" goes all the way back to the local sub-station.

However, the 800-pound gorilla is back to haunt us. As you will remember when you read the section on the explanation of *Tables A-15, A-16 & A-17*, we have a big-time problem facing us when we view the additional infrastructure that has to be built to handle the perceived increase in the usage of electricity. Whatever infrastructure is in place now must be augmented with another, parallel infrastructure 1.8 Times the size of the current one. I'm here to tell you, that is a lot of copper wire and real estate thrown at the problem. Needless to say, there needs to be some planning in the offing because that construction cannot proceed ad hoc.

## Dot #6: Renewable Resource Development

I have illustrated the massive effort that must take place to both manufacture and install any form of renewable energy acquisition hardware, be it Solar Collector Farms, Windmill arrays, Hydroelectric facilities, etc. The cost of real estate alone is staggering. The scale of need is large because the density of the energy from the Sun, wind and water is so minimal.  Read "**Parable Hour**."  To try and fragment the distribution and collection of this energy to residential and commercial roof tops is a fool's errand, primarily due to reliability issues.  See **Dot #7**.

This is a logistics and planning issue that cannot, I repeat, cannot be handled in any partisan and/or political manner.  To throw this problem at the feet of congress, or heaven forbid our current executive administrative branch, will doom it to failure.  Elon Musk is great at building EVs and flying to the moon and beyond, but this problem is of such a global scale that even he is going to have a problem wrapping his head around the solution.

The bottom line here is that we need to have level heads begin to deal with all of the intertwined issues associated with our move away from the dominance of fossil fuel as our source of energy to the use of renewables.  More on this later, but probably no film involved.

## Dot #7: Current Electricity Generator Mix

Currently, we have basically three categories of electrical generating facilities; base load, spinning reserve/base load, and peaking units.  Key to this discussion is that each and every one of these plants do not get their power from renewable sources.  They employ either nuclear or fossil fuel to generate the power and therefore are highly reliable and can be counted on to provide the necessary energy at the "meter" when called upon.

Base load plants normally are fueled by either coal or nuclear fuels and for the most part operate at 99% +/- capacity.  These are facilities whose outputs can't be ramped up and down rapidly due to either possible "melt-down" constraints, as in the nuclear arena, or simple flywheel effects caused by fueling constraints as in coal facilities.  The number of base load plants "on the line" at any moment is planned well in advance by facility engineers based on known history of usage and current, real-time usage data feedback.

Spinning reserve plants normally are either small coal plants or petroleum-fired steam plants operating at less than capacity until called upon.  These types of plants can be ramped up or down randomly without harm to the generating hardware or the grid feeder structure tying them to the grid itself.  Contrary to popular belief, the demand for electricity channeled into and out of the grid is not a smooth function of time, and in fact can change dramatically very quickly.  Fortunately, there is a substantial flywheel affect within the entire structure, augmented by large electrical capacitance naturally adjusting the power factor, such that normal fluctuations are smoothed out and the ramping functions don't have to be instantaneous.

Peaking plants are brought on-line at the start of known daily peaks in energy demand, such as during the early to late morning hours.  They are gas fired turbines or diesel engines driving generators and their power output can be adjusted very rapidly without harm.  Normally, they are kept off-line because the cost of operation is substantially greater than either the base load or spinning reserve plants.  Granted, their capital cost of construction and maintenance is less that the base load units, but the fuel costs far outweigh any gain from the amortization of the hardware itself.

The key issue here is that the output from all these plants can be relied upon almost 100%. If these plants are taken off-line permanently, or at best "moth-balled" because of the use of renewable sources, massive interruptions at the retail level can take place when those renewable sources drop off-line immediately. This is what happened within the Texas independent grid during the winter of 2020/21. That is an example of what must be considered when planning to move away from the use of fossil fuels. See **Dot #9** below.

## Dot #8: Time-of-Use and Demand-based Electricity

The consumption of electric power throughout the United States is not constant and varies both diurnally, and on a seasonal basis. Depending upon the area of the country, a particular utility may have its yearly peak occur in the summer or the winter. The daily peak normally is in the afternoon, basically due to building occupancies and the heating and air conditioning needs. A summer peaking utility is either oriented in the Southern U.S. or has natural gas or fuel oil as the predominant energy source for heating. Winter peaking utilities, as Dominion Power in VA was in the 1980s when I studied their usage, predominately have customers who use electric power as a source of heat.

As I illustrated above, peaking plants are low in capital construction costs, but consume expensive fuels at less than the efficiencies of base load plants. Therefore, the incremental cost of the "next" kWh generated on-peak is substantially greater than the "next" one generated off-peak. For the most part, common residential rate structures take into account the yearly cost of operation for the utility in question, and through the judicial process with the state corporate/utility commissions, come up with a somewhat uniform cost per kWh consumed. The site and metering costs are biased into the utility bill and incremental kWhs are priced according to the season, but not necessarily the time of day.

There is another component to the time-of-usage discussion that is quite critical to the pricing of electricity to large, commercial and industrial consumers. These customers tend to be singular and isolated with respect to the grid, as opposed to residential consumers. Also, they have a great tendency to bring on the line instantaneous demands that result when large pieces of equipment are turned on to meet industrial or commercial demands. Residential customers are viewed as a group because their demands, most notably derived from the start of air conditioners or resistive elements when heat is required, is highly diversified within neighborhoods, especially when viewed at the sub-station level. Thus, any single residence does not affect the grid demand as would say the startup of a 400 ton/horsepower centrifugal air conditioning compressor in a high-rise office building. I'm pretty sure we don't have many Clark Griswalds from National Lampoon's Christmas Vacation running around our neighborhoods shutting down the grid during the Holiday Season.

This instantaneous demand component of commercial and industrial usage is handled from a billing standpoint with a "demand meter." This type of meter "pegs" to a kW value higher than a previous value when the inrush kW exceeds that previous value. The customer is then charged monthly, or within other increments, on the registered kW demand value of the last "peg." The meter is reset after being read, and the process starts all over for the next billing period. This impact on both the customer's cost of electricity, and the grid itself, is mitigated using motor starting techniques that ramp each start up slowly, as for instance, using part-winding start methodologies. To go any further with this issue is beyond the scope of this document.

Regardless, the electric utility industry must deal with total daily/yearly usage, time-of-use demands, and the instantaneous needs of their commercial and industrial base. To attempt to seamlessly integrate renewable sources of electricity into this literal maelstrom of demand is a task that cannot be addressed flippantly.

## Dot #9: Renewable Source Problems

The fragmentation of renewable energy generation sites, especially as they relate to such areas a privately owned facilities or residences, does little to help advance progress toward the eventual elimination of a large portion of our usage of fossil fuels when used for the generation of electrical energy. In the short run, all the electrical generating facilities currently available for usage must remain as such in order to satisfy the state-mandated charters pertaining to the reliability of the grid. This means lots of spinning reserve and the subsequent loss of overall, system efficiencies.

Electrical energy generation, using renewable sources such as the Sun or the Wind, by its very nature has no built-in flywheel effect to rely upon. Thus, it is inherently un-reliable. This lack of reliability can be mitigated to some degree using the flywheel effect of battery storage techniques, and the technology that Dominion Power used in Bath County, VA. See the "**Hydroelectric**" section under "**The Sun; the Last but Not the Least.**" However, the best backup remains to be the existing generating facilities.

In order to supplant coal usage, we need to build more nuclear plants, devote enormous portions of U.S. real estate to solar and wind power generation, and get more hydroelectric flywheels in service. The key issue still is the need to have the power available when it is needed, not just when it can be generated. Backing up our power generation infrastructure with battery technology is expensive and must be addressed using rational cost/benefit analyses, not emotion or political pressure. This requires discipline and must be approached in a non-partisan manner.

## Dot #10: Time of Use, "Gas Stations," and EVs

EVs and their refueling depots are to the renewable energy generation industry as the chicken is to the egg. One of the classic statements put forth as a marketing come-on used by the EV automobile industry is that the vehicles can be recharged at night, at "home" when the generating facilities are not required to serve other, daytime needs. Surprise, surprise; not everyone lives in a single family, detached suburban house of their dreams, but they do need parking lots and parking garages.

And guess what. The Sun doesn't shine at night, the wind sort of dies down normally, so what we are left with as it pertains to a source of cheap power is the fossil fuel sector, something I thought we were trying to wean ourselves of. We must recognize that we can't ignore human nature. When we view the number of drivers who run out of gas on the highway and the lines at the gas pump during the day, I would say that we are of a culture that waits for the "near empty" light to shine before we do anything.

Probably, in the broad scheme of things, most of the recharging is going to take place during the daytime. I don't think it will take too many instances of an EV running out of battery power before recharging takes place at times of convenience. Imagine the embarrassment of running out of battery and having a gas-powered pickup truck pull up, towing a trailer with a diesel-powered generator mounted on it, for the sole purpose of recharging EV batteries for drivers in distress. Wow, what a picture that would make. In fact, I think it already has.

More on this, in depth, in **Part C**, "**What About the Automobile of the Future?**"

## Dot #11: Cherry Picking

Our current EV industry is being held up by a three-legged stool. The first leg is the massive subsidies provided to the automobile industry for the development of reasonably efficient vehicles. The second leg is the tax subsidies provided at both the State and Federal level which tend to reduce the incremental cost/benefit premium imposed due to the increased cost of technology needed to produce the EV. The third leg is the size of the cherry tree being picked dry. The EVs are being purchased because of those subsidies, granted, but the real driver is status and the demonstration of alignment with the **GND**ers. The real driver in the market should be vehicle operational economics, which currently don't exist. More on that when we discuss the **Well Running Dry**.

## Dot #12: Our Underground National Treasure

I've touched before on the concept of the fossil fuels that exist under the surface of the geographic boundaries of the United States as being a National Treasure. Well, the argument can be, and probably will be, made by the bleeding hearts of the world that what is down there really is a global treasure, not a National Treasure. If that becomes the political consensus, then our "rice-bowling" problem grows substantially. Therefore, the usage of any fossil fuel becomes a global, political problem that will dwarf the problems we are having in our pursuit of the cause of the COVID-19 pandemic. It will not take long before the next brood of "Crusader Rabbits" pops out of the woodwork, wakes up to this issue, and starts the next round of "hueing and crying."

## Dot #13: The Law of Unintended Consequences: for the Want of a Nail . . .

We're back to that pesky butterfly in San Luis Obispo and its random flight pattern. When we poke one side of the bean bag, we need to recognize that the other side of the bean bag, and all the beans therein, go somewhere. More than likely, the soldier who lost the nail from his horse's shoe never realized that the war was lost as a result. Let's pose some questions:

- Over the years, we have removed many cubic feet of petroleum and natural gas from the crust of the Earth. Essentially, this removal creates a void in the crust. Will there become a point when that void collapses and New York City becomes a Sink Hole? Maybe that's what happened to the City of Atlantis. What do you think? Given our current political/defund-the-police climate, is losing NYC a bad or a good thing? Just askin'.
- All the fossil fuels that we have consumed have in one way or another resulted in heat added to the Earth, predominately in the crust of the Earth, both solid and liquid (see **Part B**). Had we not taken advantage of that resource, $CO_2$ aside, what would be the thermal and atmospheric condition of the Earth today? If the climate models are as reliable as they are being bally hoed to be, they could tell us, but I'm not holding my breath.
- When we remove the Sun's energy that otherwise would go to provide heat into and onto the Earth's surface, and convert it into electricity, what will happen to our weather patterns?

What about those green-growies that aren't getting that energy to grow and produce trees, wheat and grass for all those little rabbits? What about having no Sun-derived energy that used to evaporate surface moisture to create clouds to block the Sun from overheating the Earth. More rain: rain where it never was before: growing the icecaps?

○ Any proliferation of wind farms is going to disturb a lot of the normal flow of air around the Earth. The "clustering" of wind farms can only go so far. A lot of elbow room needs to be given between each tower. The wind needs to recover back to "laminar" flow before it hits the next set of blades. In the transfer of heat, turbulence is good, laminar flow is bad. With wind harvesting, turbulence is bad, laminar flow is good. When we disturb the natural flow of the wind, what will be the effect on plant pollination or cloud formation or humidity content of the air? No more hurricanes or lots more hurricanes? Might be nice to know, but I'm certainly not waiting for the results from the "models."

○ The time that needs to be devoted for the recharging of EV batteries is really going to put a crimp in the style of the "Woke" crowd. One could spend a boatload of money on fast recharging batteries and the corresponding stations, but most people don't have that kind of money, and are used to filling up in maybe 10 minutes and then they are on their way. Many commutes can last hours, and span over 100 miles. I guess we are really going to have to pay close attention to that "you-gotta-recharge-NOW!" light.

○ Finally, we have the Northern part of our country to deal with come winter. A lot of that supposed efficiency is going to be thrown out the window because Mr. and Mrs. John Q. Commuter are not going to sit in traffic without some heat in the cabin. Some manufacturers are touting the use of a heat pump installed in the vehicle for both heating and cooling. Take it from an old refrigeration engineer, a heat pump won't hack it when it is below 42 °F in Minneapolis, MN. In addition, the Miami, Tampa or Fort Myers commuter is not going to sit in traffic without an air-conditioned cabin to sit in. Been there, done that. Those refrigeration compressors will drain a battery in a heartbeat.

Okay, I'm going to give you the credit of seeing the picture as it really is, as opposed to what some folks want you to see. Do shiny objects come to mind? These are questions that may, or may not, need answering, but I think they should be addressed and answered or determined to be inconsequential. So far, I've posed the problem(s) but have provided no solutions. I've told my sons, and any staff that I had reporting to me, don't come to me with just problems, bring the solutions along with you. So here you go . . .

# What are We to do Before the Well Runs Dry?

We are living in a culture that has only two components: us and fossil fuel. We, as a society, have a heretofore unheralded voracious appetite for energy of all sorts. Without energy at our fingertips, we would still be living in the dark, in caves. It is a symbiotic relationship that we must accept, understand and recognize so that our society will be able to survive in the long-term. Probably, we will have to change our "evil ways." This change will require substantial discipline to modify our habits, and that discipline must be imposed at the highest level possible. First, I'm going to talk about what we must do, and then the "how and what" we need to do to impose those cultural changes.

1.  First and foremost, we must "sell" the concept of controlling the depletion of our fossil fuels and supplanting that usage with the alternatives, which mostly comprise the use of real estate heretofore dedicated to other purposes. That real estate will be used to harvest biomass of all varieties, and generate electric power. This will be a gargantuan effort, political in nature, and of the magnitude of what FDR had to deal with to convince our pre-WW2 population that we needed to go to War to defeat Germany and Japan on the battle field. Granted, this issue is a little different than building aircraft and tanks, but if we are not successful in selling this concept, the results could be worse.

2.  The end game must be presented in a fashion that is understandable by the generally-less-than-technical members of our society. It can't be "doom and gloom," nor can it be "it's all rosy and we got it handled." We have been lied to so often by the powers-that-be here in the U. S. of A. that those same powers have a real credibility problem. A classic example is the shambles that our government, through the NIH, WHO and CDC, made of the handling of the information flow pertaining to the COVID-19 epidemic. Probably, it became a Pandemic only because China had its own agenda which didn't coincide with the best interests of the World. Does the latest debacle in Afghanistan come to mind? But I digress.

3.  History shows that without some short-cuts being taken, it takes about 20 +/- years to take a nuclear powerplant from concept to completion, including licensing, siting, environmental impact statements, etc. Currently, to my knowledge, we have no new nuclear plants "in the pipeline" so to speak, and we are decommissioning existing plants as we speak. Keep in mind that at the peak of their usage, they were only supplying about 20+/-% of our electricity. There is no question that more nuclear generation capability needs to be brought to fruition using the **Trumpian** tactics of "Operation Warp Speed." Smaller nuclear plants, like those on Nuclear powered submarines, will more than likely be employed so that their usage more resembles the coal-powered plants that they must replace.

4.  We have incredible hydroelectric resources available to us. The dams that the Bonneville Power Authority constructed on the Columbia, Snake and Clearwater rivers in Washington, Idaho and Oregon are testament to that capability. We can perform that same job in all the rivers in the United States and Canada. Quite honestly, we need to get Canada involved because as we go, so do they. They will run out of fossil fuel just as we will, so to keep them out of the loop is foolish. They have ample resources that can be utilized, but we must make sure it is a collective effort. We don't want a

situation to develop wherein Canada does nothing and has to rely upon our generation of electricity using our renewable resources.

5.  The efficiency of a fossil fuel source power plant varies as to its makeup. A coal-fired, or oil fired, steam plant is about 30% to 40% efficient, depending upon age and the design tradeoffs taken. A natural gas turbine is less efficient than the other alternatives; about 20% to 30%, best case, again depending upon the tradeoffs taken and the ambient air temperature. There is no question, as I point out below, that we need a National Plan focused on taking maximum advantage of our base-load facilities and reduce as much as possible the firing up of gas or oil-fired peaking plants. Essentially, we must "flatten" our end-use load, meaning more cultural change. Keep in mind that totally flattening the curve breeds its own problems. Much of the down time within the generating facilities is used for needed maintenance and refueling purposes. Therefore, the odds are we will never be able to take advantage of a totally flat production profile.

6.  There is no question about what will fuel our future "pedestrian" transportation needs. It will be EVs and internal combustion engines using high percentage Ethanol fuels. However, once we get beyond the "family car," we've got bigger problems. The transportation-of-goods industry, and the vehicles needed to complete its accomplishments daily, is huge, to put it mildly. In my analysis, I have not segregated the fuel usage between automobiles, trucks and trains. That is the subject of another study. However, in the short term we need to begin rational allocation of those fuels that would be used by the transportation-of-goods industry so that at least we have bread on the shelves. In the long term, we will have to create a huge, biodiesel industry because those trucks are very weight-load sensitive and they don't need to haul tons of batteries around the countryside when other options are available. In addition, railroad locomotives are motivated by diesel-driven electrical generators. The trains can't be put on a siding to rot.

7.  The environmental impact of the necessary expansion of the electrical transmission and distribution grids, and possibly the number of power plants, will be staggering. The "not-in-my-backyard" (NIMB) hue and cry will be deafening when the transmission lines and substations have to be sited in high population areas, where oh-by-the-way, the cars are needing to be re-charged. A lot of the transmission line negative NIMB cosmetics can be mitigated by siting the lines underground, which has been done in the past successfully. The surface area above ground still must be cleared and devoted to the utility easement, but at least the lines don't block the view of the Mountains. However, it comes at a recognizable, incremental installation and maintenance cost, something that must be considered within the global plan.

8.  Batteries, Windmill Generators and Solar Photovoltaic panels consume lots and lots of precious and rare-Earth minerals. These are commodities that currently are being supplied by nations that are not necessarily our allies or good friends, no matter what the Biden Administration says. Does the latest debacle in Afghanistan come to mind, because China now "has it all"? We must expand our stockpiles of these commodities and develop a plan to find deposits in our own soil that can be mined and refined economically. If this can't come to be, then we have to develop an industry that can take what we have and get it into the hands of U.S. Manufacturers so that we no longer rely on foreign sources, regardless of the cost of production.

9.  Right now, we are held hostage to the above not-so-nice players when it comes to the manufacturing of the mechanical contrivances mentioned that are required to be installed on our soil in order to generate the needed future electric power derived from the Sun. The future economics of bringing those manufacturing capabilities "back on shore" must addressed now as opposed to when it is too late, and China decides to test the Biden Administration and annex Taiwan.

10. Of greatest importance, no matter what program or programs is/are implemented, is that transparency rules and politics is completely out of the picture. We can't have this problem and its solution(s) subject to the 2-, 4-, 6- and 8-year cycles that currently drive our topsy-turvy political existence. That constitutionally imposed turbulence, if allowed to be in control, will thwart any rational execution of the "plan."

Okay, we're now down to the short strokes. The question before the house is how to get all the tasks that I have laid out accomplished. I don't think it will be easy, nor do I believe we will be 100% successful. Nor do I think anything will happen overnight. However, I think we should try. What I am going to lay out below is the ideal, but totally impractical solution, and would not have a snowballs chance in Hell of being implemented. It is authoritarianism to the $N^{th}$ degree. I put it forth here because it will be proposed by the same crowd that brought you "**Climate Change.**" Different story, same tune, and we have to recognize the tell-tale symptoms. However probably there are elements therein that might make some sense if implemented with a degree of intellectual concern.

a) We establish a congressional committee whose sole job is to create a **totally**, and I do mean **totally**, "non-partisan" cadre of experts in the following fields. One or more of these experts needs to come from each of Canada and Mexico.
   - Petroleum Engineering
   - Electrical Engineering
   - Automotive Manufacturing
   - Geophysical Exploration
   - Electric Utility Construction, with an emphasis on Nuclear
   - Residential and Commercial Building Construction
   - Environmental Impact Studies
   - Oceanography
   - Climatology
   - Global Economics
   - Mass-Marketing Techniques

b) The purpose of this body is multifold.
   - First and foremost, this body, perhaps through the help of others, must determine how to sell to the general public the concept that $CO_2$ buildup in the atmosphere, "Climate Change", and "Global Warming" are not the issue (see **Part B**). It is the depletion of our fossil fuel natural resources throughout North America. Once the sales job is complete, and most of our citizenry are informed apolitically, the rest will fall into place. And, oh-by-the-way, $CO_2$ will be reduced simply because we do something smart, if anyone cares at that point.
   - It needs to establish beyond doubt the possible, probable and likely amount of fossil fuels that are resident under our geographic boundaries. Again, this easily could, and most importantly should, involve both Mexico and Canada and their natural resources. In fact, it may be difficult to separate ownership simply delineated by political borders. At a minimum, rivers cross the borders and they are a source of hydropower.
   - Determine a methodology whereby current "ownership" of those resources can be established.
   - Establish a means of determining the in-ground value of those resources in constant dollars at a particular point in time.

- ➤ Fundamentally, get the real facts out into public view (I think we call that transparency) and let the smart people that we are make our own decisions going forward.
c) Okay, here's where the process becomes not only sticky, but perhaps totally out of the question. What I am going to put forth below is the ideal, in an ideal world, but in today's cultural and political environment, totally unworkable. It is the epitome of authoritarianism and will be dead in the water before it starts. That said, I put it forth as food for thought and perhaps a point of departure. The reason being that someone or somebody will propose a flavor of this concept and we should be prepared with alternatives.
    - ➤ Determine the legality of imposing the right of eminent domain on those resources so that our nation and others can take total control of the Natural Resource usage.
    - ➤ Develop a legislative way to compensate the owners of those resources in a way that payment only occurs when the resource is removed from the ground.
    - ➤ Put into place the skeleton of a plan that allows for the extraction of those resources in an ever-diminishing amount such that the free market, through supply and demand, will determine which end users can afford to pay the ever-increasing price, and which sectors will choose other alternatives.
        - ▪ There is a high probability that economists, once faced with the facts as presented here, will determine that currently we are paying less for our energy than we should be paying. As the old saying goes, we can only spend our money (fossil fuels) in the bank once and then it's gone.
        - ▪ This is not to say that we will avoid importing fossil fuels, nor should the importation be legislated. What we might see is that we are "leading the charge" and other suppliers will begin their own belt tightening.
        - ▪ There is no doubt this plan will, in a very short hurry, start to raise the price of all fossil fuels worldwide. Obviously, this will tend to promote conservation, but most importantly, cause us to closely examine where and how we currently are employing the use of fossil fuels.
        - ▪ Also, we may see some "rice-bowling" and resource hording take place throughout the world. The thought being that a country or entity will keep their fossil fuels off of the market, consume the resources of others, and then at a later point, when the price has escalated substantially, bring their products back on line. This of course greatly depends on how much of that entity's gross national product is the sale of fossil fuels into the world market, and how much "belt-tightening" they are willing to accept.
    - ➤ Finally, this "commission" of sorts has to develop a way of establishing a permanent body, perhaps covering the entire continent of North America, whose sole function is to control, judiciously I might add, the extraction of and dissemination of our fossil fuel resources. Let's call it the "Fossil Fuel Resource Court," or **FFRC**.
        - ▪ The members of the FFRC must be appointed for life, just like our Supreme Court. However, they are impeachable.
        - ▪ The U.S., Canadian and Mexican members must be appointed in a nonpartisan, unanimous manner by a bi-partisan group composed of the "leaders" of all political parties currently of registry.
        - ▪ Their appointment must be confirmed by a ¾ majority of both houses of congress, at least on our side. Probably, the same constraint must be imposed on the other nations involved.

- This body will have no control over the wholesale or the retail price of any fuel. They will only control what comes out of the ground and how fast. The free market will dictate the price. The point is that the ramping function is known to all and the whole process is transparent to those of us who are paying the price.
  - ➤ The overall result of implementing the FFRC program will be to stabilize the energy market. In a very short time, the dust will settle and entrepreneurs with sprout up with new and innovative concepts on how to best deal with our future as it pertains to energy utilization. There will be enough stability in the economy that investment capital will come out of the woodwork and we will prove to the world that we are the greatest country on the face of the Earth.

d) Let's look at how "dumb" Pres Biden's move to shut down the Keystone Pipeline project and open up the path for Russia to sell Natural Gas to Germany & the rest of Western Europe. In the broad scheme of things, he may have done the right thing for the wrong reason, but time will tell. There is no question that, at least in the short-term, it is to our advantage to be energy independent and boost our economy by being able to export fossil fuels. It provides a substantial degree of political independence and leverage. The other side of that coin bodes the question pertaining to some of my rantings and ravings above. What we have underground, within our sovereign boundaries, is finite. Why not use other people's resources and save ours for later dominance? Hitler lost the war, at least partially, because he ran out of fuel. Germany had/has few if any natural resources. He who has the last drop, wins.

e) The alternative to "**C**" above is to get the Federal Government out of the equation and let the free enterprise system do its magic. If we don't remove the Feds, decisions will be made, and plans implemented that have no basis to both current and long-term reality. This whole document was crafted not as an absolute problem solver or solution in and of itself. It is intended to be a point of departure and food for thought and contemplation. There is no question that we need some help, both in planning and plan execution. Come join us but do it legally. We do have laws on the books that will help lead the way.

# The Sum of the Parts

I think we have come full circle. $CO_2$ isn't bad, it's just not understood. Climate isn't changing, weather is, or maybe it isn't. That point is a "matter of taste." We have no control over the weather, so let's not waste any more time worrying about trying to change it. The Earth and the Sun are bigger than all of us combined (see **Part B**). Let's spend more time on educating our population on their use of energy and less on Racism, White Supremacy, **GND**ing, or class warfare.

We are tapping the well of a finite resource called fossil fuels at an alarming rate when viewed from 30,000 feet. When it's gone, we have no way of replenishing the supply. Therefore, as rapidly and as prudently as possible, let's replace the function of fossil fuels with renewable sources of energy; Agriculture products, Solar, Wind and Hydro Power. Let's devote that finite resource to providing us with products that otherwise cannot be economically produced using any other known alternatives. Plastics and Coke for steel production come to mind.

Finally, let's get some level, adult heads to work the problem. It must be worked outside of any, and I mean any, political or partisan environment. To bring the solution inside the D.C. swamp and subject it to our bloated, government bureaucracy will doom it to failure. In fact, the solution may never see the light of day if congress and the executive branch take charge. In essence, we must open the problem to our great cadre of entrepreneurs and let the free market rule. If we do, maybe our great-great grandchildren will still be able to buy and play with Legos®.

Before we go to **Part B**, I want to leave you with a thought that was planted in me in 1963. This is the year that Capitol Records released an album by The Kingston Trio, a well renowned Folk group of the 1960's. The first track on that old piece of vinyl was a song written by Billy Edd Wheeler, a well-known song writer from West Virginia. The title of the song was "Desert Pete." What Billy Edd Wheeler tried to tell us then is so apt today that I had to signoff this part by telling you about a portion of his philosophical outlook on life. The first and second stanza are in the first person - - a Cowboy travelling westward toward a job on a cattle run. He's crossing a small desert, thirsty, as you can image, and spies a water pump out there in the middle of nowhere. Right close to the pump is a can with a note in it. Essentially, the note gives instructions on how to take the water in a jar next to the pump, wet the new washer that Desert Pete, the writer of the note, just put in the pump, and then pump the handle like all get out. Desert Pete guarantees water will come and there will be plenty to do with as you wish. The refrain sums up the message - -

> *You've got to prime the pump. You must have faith and believe. You've got to give of yourself*
> *'fore you're worthy to receive*
> *Drink all the water you can hold. Wash your face to your feet. Leave the bottle full for others.*
> *Thank you kindly, Desert Pete"*

We must use all that fossil fuel that we have left to "prime" our energy pump. I'm old enough to have had the experience in real time of having to prime a real water pump. It ain't fun if you don't have any water to do it with. Though we can't "refill the bottle," we can certainly leave a legacy behind worthy of our resources and intelligence so that our future generations can enjoy the same glorious lifestyle that we enjoy today.

# Tables and Figures Part A

## Table A-1: Earth's Atmosphere
### Our "Dry" Atmosphere

| Atmospheric Gas | | Volume | |
|---|---|---|---|
| Name | Formula | in ppmv[2] | in % |
| Nitrogen | $N_2$ | 780,840 | 78.084 |
| Oxygen | $O_2$ | 209,460 | 20.946 |
| Argon | Ar | 9,340 | 0.934 |
| Carbon Dioxide[1] | $CO_2$ | 415.00 | 0.0415[3] |
| Neon | Ne | 18.18 | 0.001818[3] |
| Helium | He | 5.24 | 0.000524[3] |
| Methane | $CH_4$ | 1.87 | 0.000187[3] |
| Krypton | Kr | 1.14 | 0.000114[3] |
| Nitrous Oxide | $N_2O$ | 0.333 | 0.0000333[3] |
| **Not included above in our "Dry" Atmosphere** | | | |
| Water Vapor | $H_2O$ | 0 - 30,000 | 0 - 3% |

[1]As of Dec, 2020

[2]parts per million by volume

[3]Trace Gases compose 0.044143% ppmv of the atmosphere

## Table A-2: Greenhouse Effect Contribution

| Compound | Formula | Atmospheric Concentration ppmv | Contribution %[2] |
|---|---|---|---|
| Water Vapor and Clouds | $H_2O$ | Varies, up to 50,000 | 36% to 72% |
| Carbon Dioxide | $CO_2$ | 415[1] | 9% to 26% |
| Methane | $CH_4$ | 1.87 | 4% to 9% |
| Ozone | $O_3$ | 2 - 8 | 3% to 7% |

[1]As of Dec, 2020

[2]Schmidt et al.[22]. (2010) estimated that water vapor contibuted ~50%, Clouds ~25%, $CO_2$ ~20%, other gases ~5%

[3]Ozone exists almost exclusively (90%) in the stratosphere and thus may have a smaller effect than shown.

## Table A-3: Air-borne Moisture Sensitivity

| °F | 20 | 60 | 61 | 62 | 64 | 70 | 80 |
|---|---|---|---|---|---|---|---|
| lbs water per lb. dry air | 0.0023 | 0.011 | 0.0115 | 0.0119 | 0.0128 | 0.0158 | 0.0223 |
| grains water per lb. dry air | 16 | 77 | 80 | 83 | 90 | 111 | 156 |
| Decrease/Increase from 60°F | -79.2% | | 4.1% | 8.2% | 16.4% | 43.6% | 102.7% |

## Table A-4: World distribution of TPES by Source

| | | TWh[1] | Qbtu[3] | Mtoe[2] |
|---|---|---|---|---|
| Petroleum, Crude Oil | 32.0% | 51,998 | 177.4 | 4,471 |
| Coal, Peat, Shale | 27.1% | 44,036 | 150.3 | 3,786 |
| Natural Gas (Methane) | 22.2% | 36,074 | 123.1 | 3,102 |
| Biofuels and Waste | 9.5% | 15,437 | 52.7 | 1,327 |
| Nuclear | 4.9% | 7,962 | 27.2 | 685 |
| Hydro | 2.5% | 4,062 | 13.9 | 349 |
| Other (Renewables) | 1.8% | 2,925 | 10.0 | 251 |
| Total Primary Energy Supply, 2017 | 100.0% | 162,494 | 554.4 | 13,972 |
| Applying U.S. 2020 growth factor | | 166,979 | 569.7 | |

[1]Terawatthour, 1 billion KWhs
[2]Million Tonnes of Oil Equivalent
[3]Quadrillion Btu = $10^{15}$ btu

# Table A-5: TPES Conversion to Energy Carrier

| | Primary Energy Source | | Energy Carrier Transformation | | Energy Carriers |
|---|---|---|---|---|---|
| | Location | Form | | | |
| **Non-Renewable** | Planet Earth | Crude oil | Converted by | Oil Refinery | Converted to | Transportation fuels |
| | | | | | | Other Petroleum |
| | | | | | | |
| | | Coal | | Fossil Fuel Power Plant | | Heat, Mechanical Work, Electricity |
| | | Natural Gas | | Fossil Fuel Power Plant | | Heat, Mechanical Work, Electricity |
| | | | | Pipeline Transportation | | Living Space Comfort Conditioning |
| | | Natural Uranium | | Nuclear Power Plant | | Heat (Enthalpy), Electricity |
| | | Natural Thorium | | Thorium Breeder Reactor | | Heat (Enthalpy), Electricity |
| **Renewable** | Sun | Electromagnetic Waves | Converted by | Photovoltaics | Converted to | Electricity |
| | | | | Solar Furnace | | Heat (Enthalpy) |
| | | Wind | | Windmills, Wave Power Facility | | Mechanical Work, Electricity |
| | | Gravitational Water Flow | | Hydroelectric, Tidal Facility | | Mechanical Work, Electricity |
| | | Biomass | | Biomass Power Plant | | Heat (Enthalpy), Electricity |
| | Planet Earth | Geothermal | | Geothermal Power Facility | | Heat (Enthalpy), Electricity |

## Table A-6
## U. S. Energy Consumption by Source, 2020

|  | Qbtu | TWh | % |
|---|---|---|---|
| **Fossil Fuel** |  |  |  |
| Coal | 9.18 | 2,691 | 9.9% |
| Natural Gas | 31.54 | 9,245 | 33.9% |
| Petroleum | 32.23 | 9,446 | 34.7% |
| **Sub-Total** | 72.94 | 21,377 | 78.5% |
| **Nuclear** | 8.25 | 2,417 | 8.9% |
| **Renewable Energy** |  |  |  |
| Hydroelectric | 2.59 | 760 | 2.8% |
| Geothermal | 0.21 | 63 | 0.2% |
| Solar | 1.25 | 365 | 1.3% |
| Wind | 3.01 | 881 | 3.2% |
| Biomass | 4.53 | 1,328 | 4.9% |
| **Sub-Total** | 11.59 | 3,397 | 12.5% |
| **Total** | **92.94** | **27,240** | **100.0%** |

[1]**Note**: There exists a 4% difference in heat content data for Petroleum Usage in the U. S. when comparing Tables 1.3 and 3.6, EIA Monthly Energy Report, April 2021. This reason for this difference is not readily apparent in the literature. A possible explanation could focus on the ascribing of heat value in the various sectors of Petroleum usage, given the different qualities of raw petroleum consumed in each sub-sector. Rounding errors cannot account for all of this discrepancy.

*Note: All data is presented in rounded format. The summation of related, rounded data due to rounding techniques employed by the author will not always be equal to the final sum presented. A detailed explanation of the data in each of the tables 6 – 8 is given in the section* **"Explanations of Tables A-6 – A-8."**

## Table A-7
## U.S. Energy Consumption for Electricity Generation, 2020
### 38.5% of Total Energy Consumption

|  | Qbtu | TWh | % Elec | % Total |
|---|---|---|---|---|
| **Fossil Fuels** |  |  |  |  |
| Coal | 8.231 | 2,412 | 23.0% | 8.9% |
| Natural Gas | 11.972 | 3,509 | 33.5% | 12.9% |
| Petroleum | 0.18 | 53 | 0.5% | 0.2% |
| **Sub-Total** | 20.382 | 5,974 | 57.0% | 21.9% |
| **Nuclear** | 8.248 | 2,417 | 23.1% | 8.9% |
| **Renewable Energy** |  |  |  |  |
| Hydroelectric | 2.581 | 756 | 7.2% | 2.8% |
| Geothermal | 0.147 | 43 | 0.4% | 0.2% |
| Solar | 0.802 | 235 | 2.2% | 0.9% |
| Wind | 2.998 | 879 | 8.4% | 3.2% |
| Biomass | 0.424 | 124 | 1.2% | 0.5% |
| **Sub-Total** | 6.952 | 2,038 | 19.4% | 7.5% |
| **Total** | **35.744** | **10,476** | **100.0%** | **38.5%** |

## Table A-8
### U.S. Electrical Energy Distribution by Sector, 2020

|  | Qbtu | TWh | % Retail Sales | % of Total Energy Generation Consumption | U. S. Consumption |
|---|---|---|---|---|---|
| Residential Retail Sales | 4.988 | 1,462 | 39.9% | 14.0% | 5.4% |
| Commercial Retail Sales | 4.353 | 1,276 | 34.8% | 12.2% | 4.7% |
| Industrial Retail Sales | 3.137 | 919 | 25.1% | 8.8% | 3.4% |
| Transportation Retail Sales | 0.022 | 6 | 0.18% | 0.06% | 0.02% |
| Total Retail Sales | 12.500 | 3,664 | 100.00% | 34.97% | 13.45% |
| Generation Losses | 23.243 | 6,812 |  | 65.03% | 25.01% |
| Total | 35.74 | 10,476 |  | 100.00% | 38.46% |

## Table A-9
### U.S. Petroleum Consumption by End Use, 2020

|  | Thousands of Barrels per Day[1] | Qbtu per Year | TWh per year | Most Probable Reduction | Net TWh | Ideal Case Possible Reduction | Net TWh |
|---|---|---|---|---|---|---|---|
| Asphalt and Road Oil | 342 | 0.831 | 243.6 | 0.00% | 242.8 | 0.00% | 242.8 |
| Aviation Gasoline | 11 | 0.020 | 5.9 | 20.00% | 4.8 | 75.00% | 1.5 |
| Distillate Fuel Oil | 3,776 | 7.956 | 2,331.8 | 11.78% | 2,056.5 | 55.93% | 1,027.3 |
| HGL[2] - Propane | 812 | 1.142 | 334.7 | 21.08% | 263.6 | 61.62% | 128.2 |
| HGL[2] - Propylene | 277 | 0.389 | 114.0 | 0.00% | 113.6 | 0.00% | 113.6 |
| HGL[2] - Other | 2,108 | 2.387 | 699.6 | 0.00% | 755.2 | 0.00% | 755.2 |
| Jet Fuel | 1,078 | 2.237 | 655.6 | 0.00% | 653.9 | 10.00% | 588.5 |
| Kerosene | 8 | 0.016 | 4.7 | 25.00% | 3.6 | 68.75% | 1.5 |
| Lubricants | 100 | 0.223 | 65.4 | 0.00% | 65.5 | 0.00% | 65.5 |
| Motor Gasoline | 8,034 | 14.855 | 4,353.8 | 38.89% | 2,653.5 | 100.00% | 0.0 |
| Petroleum Coke | 260 | 0.582 | 170.6 | 15.00% | 166.7 | 15.00% | 166.7 |
| Residual Fuel Oil | 217 | 0.499 | 146.2 | 18.25% | 119.3 | 50.00% | 73.0 |
| Other[3] | 1,099 | 2.396 | 702.2 | 0.00% | 684.8 | 0.00% | 684.8 |
| Total | 18,122 | 33.533 | 9,850.2 | 20.98% | 7,783.8 | 60.93% | 3,848.6 |

[1]Table 3.5, Petroleum Products Supplied by Type, U.S. EIA/Monthly Energy Review April 2021
[2]HGL - Hydrocarbon Gas Liquids
[3]Petrochemical Feed stocks, Waxes, etc.

# Table A-10
## U.S. Petroleum Consumption by End Use, 2020

|  | Thousands of Barrels per Day[1] | Qbtu per Year | TWh per year | Most Probable Reduction | Net TWh | Ideal Case Possible Reduction | Net TWh |
|---|---|---|---|---|---|---|---|
| Residential | 566 | 1.047 | 273.4 | 30.0% | 191.37 | 75.0% | 68.34 |
| Commercial | 428 | 0.792 | 226.2 | 19.4% | 182.22 | 79.7% | 45.82 |
| Industrial | 5,082 | 9.403 | 2506.9 | 2.1% | 2,453.49 | 15.3% | 2,124.61 |
| Transportation | 11,964 | 22.136 | 6789.0 | 27.7% | 4,910.22 | 76.7% | 1,580.32 |
| Electric Power | 84 | 0.155 | 54.6 | 14.8% | 46.53 | 46.0% | 29.47 |
| **Total** | **18,122** | **33.533** | **9,850.2** | **20.98%** | **7,783.8** | **60.93%** | **3,848.6** |

# Table A-11
## U.S. Natutral Gas Consumption by Sector, 2020

|  | Billion Cubic Feet per Year | Qbtu per Year | TWh per year | Most Probable Reduction | Net TWh | Ideal Case Possible Reduction | Net TWh |
|---|---|---|---|---|---|---|---|
| Lease and Plant Fuel | 1,829 | 1.897 | 555.9 | 5.0% | 528.09 | 20.0% | 444.71 |
| Pipeline & Distribution | 926 | 0.960 | 281.4 | 5.0% | 267.36 | 20.0% | 225.15 |
| Residential | 4,648 | 4.820 | 1412.7 | 30.0% | 988.86 | 85.0% | 211.90 |
| Commercial | 3,147 | 3.263 | 956.5 | 30.0% | 669.52 | 85.0% | 143.47 |
| Industrial CHP[1] | 1,367 | 1.418 | 415.5 | 30.0% | 290.83 | 70.0% | 124.64 |
| Industrial Non-CHP[1] | 6,890 | 7.145 | 2094.1 | 0.0% | 2,094.06 | 0.0% | 2,094.06 |
| Vehicle Fuel | 59 | 0.061 | 17.9 | 30.0% | 12.55 | 90.0% | 1.79 |
| Electric Power | 11,616 | 12.046 | 3530.4 | 10.0% | 3,177.38 | 50.0% | 1,765.21 |
| **Total** | **30,482** | **31.610** | **9,264.3** | **13.3%** | **8,028.65** | **45.9%** | **5,010.93** |

[1]Combined Heat and Power

## Table A-12
## U.S. Coal Consumption by Sector, 2020

| | Thousand Short Tons per Year[2] | Qbtu per Year[3] | TWh per year | Most Probable Reduction | Net TWh | Ideal Case Possible Reduction | Net TWh |
|---|---|---|---|---|---|---|---|
| **Commercial** | **793** | **0.0145** | **4.3** | **11.6%** | **3.8** | **28.9%** | **3.02** |
| CHP[1] | 459 | 0.0084 | 2.5 | 20.0% | 2.0 | 50.0% | 1.23 |
| Non-CHP[1] | 334 | 0.0061 | 1.8 | 0.0% | 1.8 | 0.0% | 1.79 |
| **Industrial** | **39,998** | **0.9362** | **274.4** | **4.3%** | **262.5** | **10.8%** | **244.80** |
| Coke Plants | 14,414 | 0.4139 | 121.3 | 0.0% | 121.3 | 0.0% | 121.32 |
| CHP[1] | 9,890 | 0.2019 | 59.2 | 20.0% | 47.3 | 50.0% | 29.59 |
| Non-CHP[1] | 15,694 | 0.3204 | 93.9 | 0.0% | 93.9 | 0.0% | 93.90 |
| **Electric Power** | **436,524** | **8.2311** | **2412.4** | **25.0%** | **1809.3** | **100.0%** | **0.00** |
| Total, All | 477,315 | 9.182 | 2,691 | 22.9% | 2,076 | 88.1% | 248 |
| Total, Non-CHP[1] | 16,028 | 0.326 | 95.7 | 0.0% | 95.7 | 0.0% | 95.7 |

[1]Combined Heat and Power
[2]Table 6.2, Coal Consumption by Sector, U.S. EIA/Monthly Energy Report April 2021
[3]Table A5, Approximate Heat Content of Coal and Coal Coke,
 U.S. EIA/Monthly Energy Review April 2021

## Table A-13
## Summary of Reduction in Fossil Fuel Usage

| | Most Probable Reduction TWh / year | | | Ideal Case Possible Reduction TWh / year | | |
|---|---|---|---|---|---|---|
| | Fuel | | Heat | Fuel | | Heat |
| | Electricity | Transport | | Electricity | Transport | |
| Petroleum | 8.10 | 1878.81 | 179.43 | 25.15 | 5208.71 | 767.72 |
| Natural Gas | 477.68 | 5.38 | 752.60 | 2056.04 | 16.14 | 2181.21 |
| Coal | 615.43 | 0.00 | 0.00 | 2443.21 | 0.00 | 0.00 |
| Total | 1,101.21 | 1,884.19 | 932.03 | 4,524.40 | 5,224.85 | 2,948.93 |

## Table A-14
## The Probable Bandwidth of the Electricity Generation Increase
### Reduction of Petroleum used for transportation, TWh/year

|  | Trillion Miles per year @ MPG[a] | | TWh per year @MPG and 3.63 miles/kWh, at the vehicle | | TWh per year @MPG and 3.63 miles/kWh, at the source | |
|---|---|---|---|---|---|---|
|  | 25.00 | 10.00 | 25.00 | 10.00 | 25.00 | 10.00 |
| Most Probable | **1878.81** | **1.42** | 0.57 | **392** | 157 | **458** | 183 |
| Ideal Case | 5208.71 | 3.95 | 1.58 | 1,086 | 435 | 1,270 | 508 |

[a]Note: The conversion factor used is 33 kWh per gallon of Automotive Gasoline, which is the generally accepted value. Marks Mechanical Engineering Handbook states a bandwidth of 37 kWh to 35 kWh using the high and low heat values of the gasoline alone. The reduction to 33 kWh is the result of the blending of refined fuel and other flammable hydrocarbons such as Ethanol at the pump.

## Table A-15
## Total Electricity Replacement at the Source

| Retail Electricity Replaced @ | Most Probable Reduction TWh / year | | | Ideal Case Possible Reduction TWh / year | | |
|---|---|---|---|---|---|---|
|  | Fuel | | Heat | Fuel | | Heat |
|  | Electricity | Transport |  | Electricity | Transport |  |
| 25 MPG | 450.46 | 459.25 | 1089.29 | 1850.74 | 1273.47 | 3446.53 |
| 10 MPG | 450.46 | 184.45 | 1089.29 | 1850.74 | 511.65 | 3446.53 |
| Total 25 MPG | 1999.00 | | | 6570.74 | | |
| Total 10 MPG | 1724.20 | | | 5808.92 | | |

## Table A-16
### Renewable Energy Source Real Estate Requirements

| | MPG | New TWh /year | % of Current | Raw Area Required Square Miles | | | | | | Wind[a] Density |
|---|---|---|---|---|---|---|---|---|---|---|
| | | | | Solar Panel Real Estate Density | | | | | | |
| | | | | 100% | | 50% | | 25% | | |
| Capacity Factor | | | | 12% | 40% | 12% | 40% | 12% | 40% | 25% |
| Most Probable | 25 | 1,999 | 54.6% | 9,046 | 2,714 | 18,092 | 5,428 | 36,184 | 10,855 | 11,748 |
| Ideal Case | 25 | 6,571 | 179.4% | 29,735 | 8,920 | 59,469 | 17,841 | 118,939 | 35,682 | 38,610 |

[a]Wind currently is believed to have 0.3 MW of "tapable" energy available per hectare. 1 hectare is 10,000 square meters, and there are 259 hectares per square mile. Thus, there is 77.7 MW of "tapable" wind energy available per square mile.

## Table A-17
### Representative State-Wide Land Areas

| | Miles² | | | Miles² |
|---|---|---|---|---|
| Delaware | 1,949 | | Indiana | 35,826 |
| Connecticut | 4,842 | | Iowa | 55,857 |
| Maryland | 9,707 | | North Dakota | 69,001 |
| West Virginia | 24,038 | | Arizona | 113,594 |
| South Carolina | 30,031 | | California | 155,779 |

# Figure A-1

## U.S. Energy Consumption by Source and Sector, 2020

Sources: U.S. Energy Information Administration
*Monthly Energy Report* (April 2021, Table 1.3)

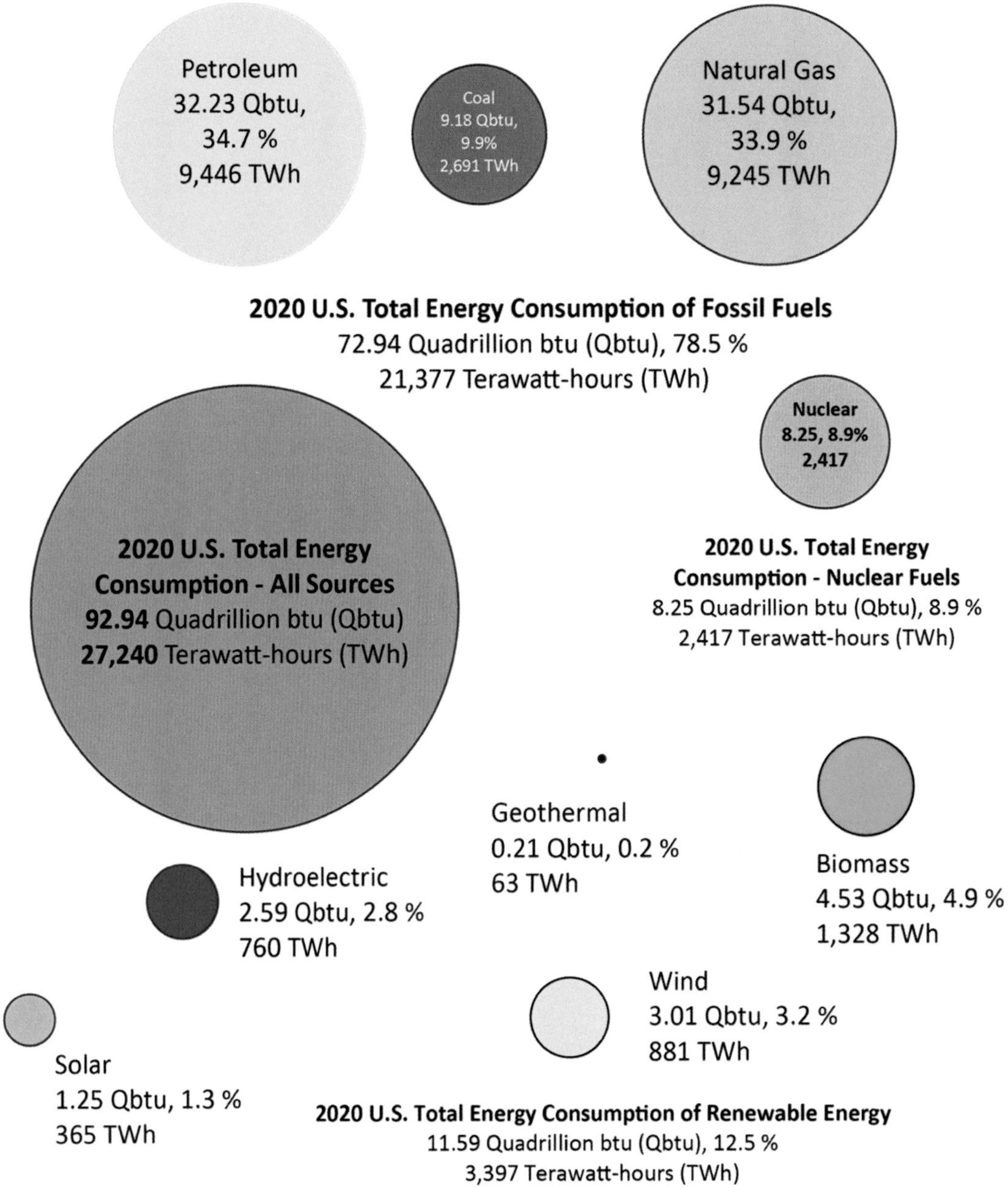

Petroleum
32.23 Qbtu,
34.7 %
9,446 TWh

Coal
9.18 Qbtu,
9.9%
2,691 TWh

Natural Gas
31.54 Qbtu,
33.9 %
9,245 TWh

**2020 U.S. Total Energy Consumption of Fossil Fuels**
72.94 Quadrillion btu (Qbtu), 78.5 %
21,377 Terawatt-hours (TWh)

Nuclear
8.25, 8.9%
2,417

**2020 U.S. Total Energy
Consumption - All Sources
92.94** Quadrillion btu (Qbtu)
**27,240** Terawatt-hours (TWh)

**2020 U.S. Total Energy
Consumption - Nuclear Fuels**
8.25 Quadrillion btu (Qbtu), 8.9 %
2,417 Terawatt-hours (TWh)

Geothermal
0.21 Qbtu, 0.2 %
63 TWh

Biomass
4.53 Qbtu, 4.9 %
1,328 TWh

Hydroelectric
2.59 Qbtu, 2.8 %
760 TWh

Wind
3.01 Qbtu, 3.2 %
881 TWh

Solar
1.25 Qbtu, 1.3 %
365 TWh

**2020 U.S. Total Energy Consumption of Renewable Energy**
11.59 Quadrillion btu (Qbtu), 12.5 %
3,397 Terawatt-hours (TWh)

# Figure A-2  U.S. Electric Power Sector Energy Consumption and Distribution, 2020

Sources: U.S. Energy Information Administration, *Monthly Energy Report* (April 2021, Tables 2.1 – 2.6)
% Shown in Energy Consumption section = % of Total Consumption for Electrical Production

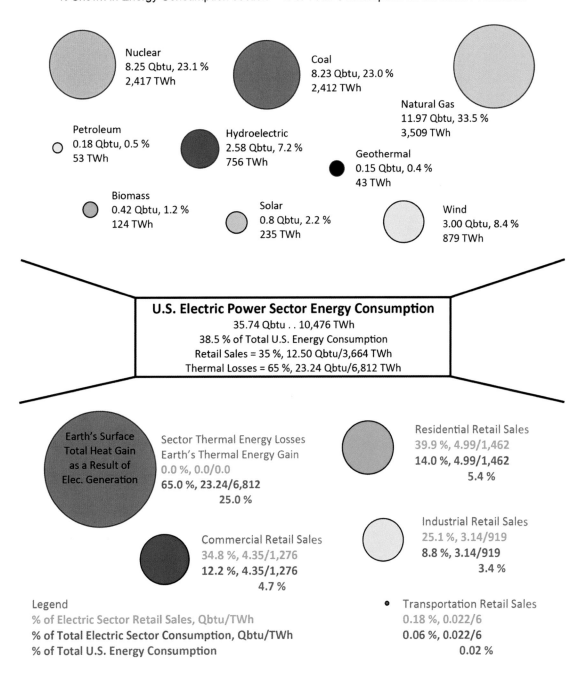

Nuclear
8.25 Qbtu, 23.1 %
2,417 TWh

Coal
8.23 Qbtu, 23.0 %
2,412 TWh

Natural Gas
11.97 Qbtu, 33.5 %
3,509 TWh

Petroleum
0.18 Qbtu, 0.5 %
53 TWh

Hydroelectric
2.58 Qbtu, 7.2 %
756 TWh

Geothermal
0.15 Qbtu, 0.4 %
43 TWh

Biomass
0.42 Qbtu, 1.2 %
124 TWh

Solar
0.8 Qbtu, 2.2 %
235 TWh

Wind
3.00 Qbtu, 8.4 %
879 TWh

**U.S. Electric Power Sector Energy Consumption**
35.74 Qbtu . . 10,476 TWh
38.5 % of Total U.S. Energy Consumption
Retail Sales = 35 %, 12.50 Qbtu/3,664 TWh
Thermal Losses = 65 %, 23.24 Qbtu/6,812 TWh

Earth's Surface Total Heat Gain as a Result of Elec. Generation

Sector Thermal Energy Losses
Earth's Thermal Energy Gain
0.0 %, 0.0/0.0
**65.0 %, 23.24/6,812**
**25.0 %**

Residential Retail Sales
39.9 %, 4.99/1,462
**14.0 %, 4.99/1,462**
**5.4 %**

Industrial Retail Sales
25.1 %, 3.14/919
**8.8 %, 3.14/919**
**3.4 %**

Commercial Retail Sales
34.8 %, 4.35/1,276
**12.2 %, 4.35/1,276**
**4.7 %**

Transportation Retail Sales
0.18 %, 0.022/6
**0.06 %, 0.022/6**
**0.02 %**

Legend
% of Electric Sector Retail Sales, Qbtu/TWh
**% of Total Electric Sector Consumption, Qbtu/TWh**
**% of Total U.S. Energy Consumption**

©2021 Richard D. Jones

# Figure A-3  U.S. Petroleum Consumption by Sector, 2020

Sources: U.S. Energy Information Administration, *Monthly Energy Report* (April 2021, Tables 3.5, 3.6, 3.7a-c, 3.8a-c)
**Note** that the scale has changed from Figures A-1 & A-2 in order to provide visual emphasis

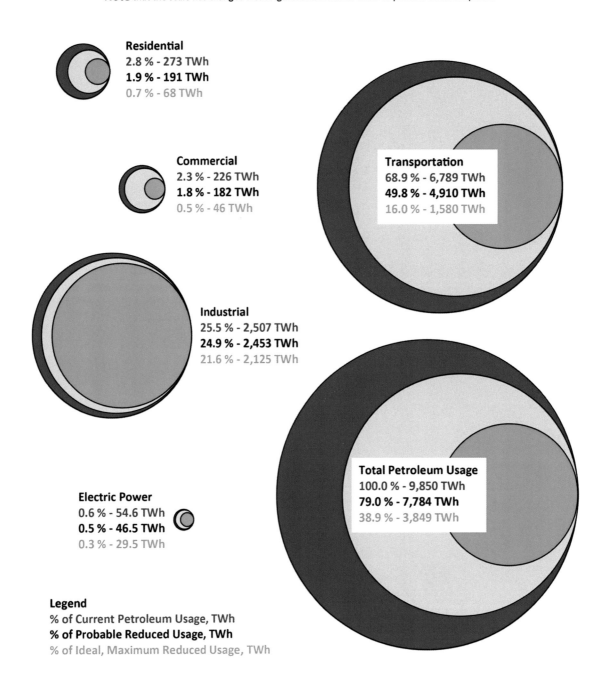

**Residential**
2.8 % - 273 TWh
**1.9 % - 191 TWh**
0.7 % - 68 TWh

**Commercial**
2.3 % - 226 TWh
**1.8 % - 182 TWh**
0.5 % - 46 TWh

**Transportation**
68.9 % - 6,789 TWh
**49.8 % - 4,910 TWh**
16.0 % - 1,580 TWh

**Industrial**
25.5 % - 2,507 TWh
**24.9 % - 2,453 TWh**
21.6 % - 2,125 TWh

**Electric Power**
0.6 % - 54.6 TWh
**0.5 % - 46.5 TWh**
0.3 % - 29.5 TWh

**Total Petroleum Usage**
100.0 % - 9,850 TWh
**79.0 % - 7,784 TWh**
38.9 % - 3,849 TWh

**Legend**
% of Current Petroleum Usage, TWh
**% of Probable Reduced Usage, TWh**
% of Ideal, Maximum Reduced Usage, TWh

# Figure A-4  U.S. Natural Gas Consumption by Sector, 2020

Sources: U.S. Energy Information Administration, *Monthly Energy Report* (April 2021, Tables 2, 4.3, A4)
CHP[1]: Combined Heat and Power.  It is assumed that none of the energy is waste product.
**Note** that the scale has changed from Figures A-1 & A-2 in order to provide visual emphasis

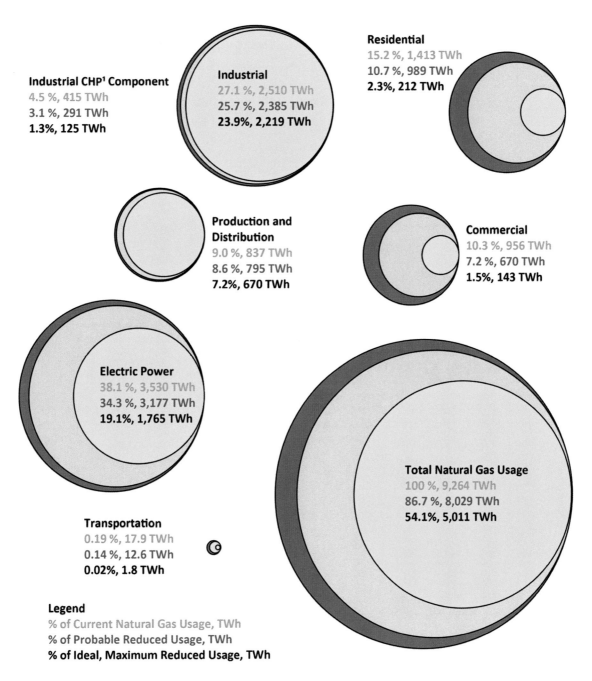

**Industrial CHP[1] Component**
4.5 %, 415 TWh
3.1 %, 291 TWh
**1.3%, 125 TWh**

**Industrial**
27.1 %, 2,510 TWh
25.7 %, 2,385 TWh
**23.9%, 2,219 TWh**

**Residential**
15.2 %, 1,413 TWh
10.7 %, 989 TWh
**2.3%, 212 TWh**

**Production and Distribution**
9.0 %, 837 TWh
8.6 %, 795 TWh
**7.2%, 670 TWh**

**Commercial**
10.3 %, 956 TWh
7.2 %, 670 TWh
**1.5%, 143 TWh**

**Electric Power**
38.1 %, 3,530 TWh
34.3 %, 3,177 TWh
**19.1%, 1,765 TWh**

**Total Natural Gas Usage**
100 %, 9,264 TWh
86.7 %, 8,029 TWh
**54.1%, 5,011 TWh**

**Transportation**
0.19 %, 17.9 TWh
0.14 %, 12.6 TWh
**0.02%, 1.8 TWh**

**Legend**
% of Current Natural Gas Usage, TWh
% of Probable Reduced Usage, TWh
**% of Ideal, Maximum Reduced Usage, TWh**

©2021 Richard D. Jones

# Figure A-5   U.S. Coal Consumption by Sector, 2020

Sources: U.S. Energy Information Administration, *Monthly Energy Report* (April 2021, Tables 6.2, A5)
CHP[1]: Combined Heat and Power.  It is assumed that none of the energy is waste product.
**Note** that the scale has changed from Figures A-1 & A-2 in order to provide visual emphasis

**Commercial**
**0.16 %, 4.25 TWh**
0.14 %, 3.76 TWh
**0.11%, 3.02 TWh**
0.07%, 1.79 TWh, non-CHP[1]
Note that the scale is changed substantially

**Industrial**
**10.2 %, 274 TWh**
9.8 %, 263 TWh
**9.1%, 245 TWh**
8.0%, 215 TWh, non-CHP[1]

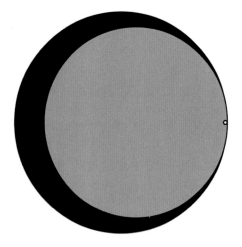

**Electric Power**
**89.6 %, 2,412 TWh**
67.2 %, 1,809 TWh
**0.0%, 0 TWh**

**Total Coal Usage**
**100 %, 2,691 TWh**
77.1 %, 2,076 TWh
**9.21%, 248 TWh**
8.06%, 217 TWh, non-CHP[1]

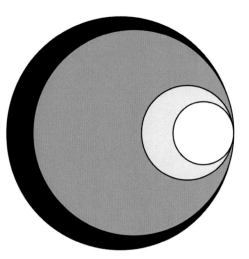

**Legend**
**% of Current Coal Usage, TWh**
% of Probable Reduced Usage, TWh
**% of Ideal, Maximum Reduced Usage, TWh**
% of non-CHP Usage, TWh

©2021 Richard D. Jones

# The Shiny Object in the Room: Part B

# Introduction

What I will put forth in this document is a simple application of science with a little common sense thrown into the mix. My goal is to make sure that you are not subjected to another **Zohnerism (see Section 10 below),** which I believe has taken place thanks to the IPCC. The issue is twofold. We are running out of a finite resource called fossil fuels, so let's employ some good asset stewardship and use what's left wisely. Second, we can't let the shiny object called "climate change" taint our thinking and critical decision making. We really have very little control over the weather and most importantly who is to say the changing weather patterns are going to be harmful. Gypsies maybe? There is a distinct possibility we might be better off. Regardless, let's worry about what we can control rather than what we think we might be able to control. Bring out that crystal ball and let's get rollin'.

First of all, I think we need to recognize, and accept, the fact that the scientific data generated by the **National Aeronautics and Space Administration (NASA)** via the **Clouds and the Earth's Radiant Energy System (CERES)** project, which was launched in November of 1997[30] has pretty well determined that our Earth has increased in temperature. Some refer to this warming trend as "Climate Change," others use the term "Global Warming." Per the results of the NASA study to date, there is no question but that Spaceship Earth, the "globe," is warming. Conventional wisdom says that from the 1850 to 1900 baseline average, the temperature of the globe has increased about 1.2 °C or 2.2 °F through 2020[31]. That said, to employ the term "Climate Change" while referencing the same data and information seems a stretch to me. What we are going to talk about here is:

Section 1: What is the difference between "Climate" and "Weather?"
Section 2: What are the elements of our tenure here on Earth that influence our micro and macro weather?
Section 3: The human influence on weather
Section 4: What is the real connection between the weather and GHGs?
Section 5: What is the history of the world's climate and how severe were the "changes?"
Section 6: Why are we warm and not cold?
Section 7: Will the $CO_2$ "Problem" take care of itself?
Section 8: The results of the model.
Section 9: Who profits by making us both aware of, and afraid of, any buildup of GHGs?
Section 10: We've been "42'd"
Section 11: What should we really focus on in order to maintain our glorious lifestyle?

# 1. Is it Climate or Weather?

That is a really, great question because the two terms are oft confused.

## Weather

". . the state of the atmosphere with respect to heat or cold, wetness or dryness, calm or storm, clearness or cloudiness"[32] (*noun, Merriam-Webster*)

## Climate

". . the average course or condition of the weather at a place usually over a period of years as exhibited by temperature, wind velocity, and precipitation"[33] (*noun, Merriam-Webster*)

So, essentially, weather really is the "here and now," while climate is an amalgam of a whole bunch of "weathers" that are part of history, and hopefully somewhat well documented. **Old Timers** always complain about the weather, followed by the fact that we can't do anything about it. Sage observation, if I do say so myself. On the flip side, no person completely in control of their faculties complains about the "climate" in real-time, knowing that it really is history, and by definition, nothing can be done to change it. In other words, weather changes, climate doesn't.

As *Merriam-Webster* so aptly states, weather is many things, whether we like it or not. Even while I was a young whippersnapper during my formative years, I questioned many things having to do with weather and my surrounding ambient in general. I didn't like the bitter cold when I had to deliver papers in the early AM. I didn't like the rain when I wanted to play golf. Loved the snow on school days, but not while I was learning to drive a car. You get the picture. That said, it was not until now, during my waning years, that a mis-guided hue and cry erupted about "Climate Change" and the evils of $CO_2$ emissions.

What became obvious to me recently, while our society is navigating its way through the morass of the above narrative, is that we have failed to instill in our educational system the need to understand the science behind those elements that govern our daily, monthly and yearly comfort. I guess I was lucky because I had great science teachers in all levels of my education who taught me to be curious and not accept "conventional" thought as gospel. OBTW, this was during the time when real science was being taught, not pseudo-science. They taught me to seek truth, using rigorous scientific disciplines and just plain logic. If it didn't look like a duck, walk like a duck or quack like a duck, no matter what anyone else said, it probably wasn't a duck. The real issue at hand therefore was what was it if not a duck?

As we go forward with this tome, you might call it "The Great Duck Hunt."  I had a feeling a long time ago, when the $CO_2$ - Climate Change – Global Warming hue and cry erupted, and assuming that the globe was really warming, per se, that, a) it was far more complex an issue than just $CO_2$ emissions, b) the Earth itself is so huge and its thermal mass is so big, burning coal to produce electricity can't be the only driver changing our world, and c) who started the controversy and who wins and who loses when the dust settles, if it ever will?  Let's proceed to find fact and ask some real good questions.

# 2. What Drives the Weather?

The answer is **HEAT.** It doesn't get much simpler than that. We have three basic sources of heat, at the surface, here on spaceship Earth. The Sun, the "Human Factor" (the combustion of fossil fuels, nuclear fission employed in the production of electricity, human-body derived energy), and that huge nuclear furnace at the center of the Earth. An overwhelming percentage of the electricity and fossil fuel consumed, regardless of geographic locale, eventually ends up as heat accruing into the atmosphere, but eventually almost 100% into or onto the multitude of surfaces within or on the Earth. This is kinda like a law of physics/nature. Thus, going forward, we will assume that all consumption of Fossil Fuels and Nuclear Energy results in Heat added to the surface. Granted, it is not instantaneous, but eventually the result of decay or combustion will add Heat to the air which will accrue quite quickly to the surface of the Earth.

*Figure B-1*: **Earth's Acquisition of Heat** demonstrates how the Earth is heated. Of course, the primary, and major, source of heat is the Sun. Conventional wisdom, coupled with real science, has shown that on the average we will see 1361 W/m$^2$ (SI units, International System of Units) falling on our outer atmosphere; the "Solar Constant[34]" or solar insolence. In IP (inch-pound) terms, this is **431** btu/ft$^2$/hour. As we will see later, this can vary substantially, and that variance in fact has a rather large impact on our climate, past and present. But that is less than half of the story. That much energy does not reach the surface of the Earth. There is a substantial portion of what falls outside of our atmosphere that is either reflected or absorbed in the Stratosphere before it has a chance to reach the surface, where it falls on water or the solid crust of the Earth. This "striping" of the Sun's energy is termed as the Albedo Effect[35]. In the absolute best case, about **317** btu/ft$^2$/hour +/- makes it way, through the Tropopause, to the surface for our use, at least in the "habitable" latitudes. See "**Parable Hour**" in **Part A**. Now, this is not guaranteed. Lots of stuff gets in the way, like GHGs, other atmospheric gases, dust, water vapor, clouds, etc. You get the picture. Most importantly, this value is only valid if the "surface" is exactly perpendicular to the Sun's rays, which of course happens only at one, teeny-tiny location on the surface at any moment in time.

Again, looking at *Figure B-1*, you will notice there is the second source of heat energy impinging on the surface, mostly over land I might add. This is the human factor that I alluded to above. I touch on this issue here, but we will pursue in much more depth in **Section 3** when we talk about the Earth's energy balance.

One thing we need to keep in mind that I mentioned before and that is that weather is local. This really means that whatever is happening "locally" probably is influencing our current weather. Remember the butterfly in San Luis Obispo and its flight path. On an hour-by-hour or day-to-day basis, we must be careful about assuming that what happens in London, England now affects our rain forecast in the morning. The only question is what is "local?" We'll talk about that later.

The solar energy that makes it through the outer atmosphere basically is composed of three different categories, or groups of wave lengths, which help us in numerous ways[36]. First is the infrared (IR), or near infrared, region which has wavelengths > 700 nanometers. The visible region, 400 to 700 nanometers, is our "light bulb" and the main source of energy for photosynthesis.  The ultraviolet (UV) region, < 400 nanometers is great for suntans, but lousy for skin cancer.  Most, but not all, of the harmful UV is blocked by the ozone layer which is much higher than the illustration in *Figure B-1* and has little if anything to do with providing heat to the surface of the Earth.

# 3. The Human Factor

I brought up the "**Human Factor**" (HF) for a reason, because it probably has a greater influence on our local weather patterns than we have been led to believe. I am going to lump three basic categories into the HF. A tiny percentage of these influences existed 10,000 years ago, and only really began to surface between 1850 and 1900 AD. The first is the consumption of fossil fuels, either as a source of heat, per se, or as a source of power used to generate steam and electricity. The second is the advent of the use of nuclear energy, again as a source of heat used to generate electricity. Finally, through our human and animal populations. Albeit, humans are sorta new, but the animals have been here for a while.

The aerobic respiration process, through our human metabolic function, results in a quite identifiable source of heat. I have identified in **Part A** the World energy consumption, TPES, and the U. S. energy consumption in all categories. The human body radiates, convects and conducts heat continuously to its surroundings. Obviously, this rejection of heat ends up on the surface of the Earth, concentrated in populated locales. More on that degree of concentration later.

The ASHRAE Guide has an extensive table of values used by HVAC engineers to determine the air conditioning needs, in confined areas such as office buildings, gymnasiums, and residential units. The 1997 ASHRAE Fundamentals Handbook, pg. 28.8, Table 3, lists typical values for both sensible and latent heat gains/depositions for various indoor activities. These are the values used by engineers, coupled with projected occupancy rates, to determine the level of air conditioning needed to maintain a reasonable comfort level in any occupied space. OBTW, we as humans produce this same amount of heat energy no matter whether we are indoors or outdoors. *Table B-1*: **Human Energy Contribution** illustrates some typical values.

Sensible heat is that which we term as energy used to raise the temperature of our living environment directly, like radiation from the surface of our bodies. Latent heat is in the form of water vapor. Essentially, it is the amount of energy we expend to evaporate excess moisture from our body. We produce or take in liquid water, some of which is expelled as water. The rest is expelled through respiration or perspiration. Water requires ~1,000 btu/lb. (about 1 pint) to change its state from a liquid to a vapor. In the air conditioning world, we must expend that same amount of energy per lb. to remove the vapor from the air as water (condensate). This water is then expelled from the air conditioning hardware back into the environment. There is a double whammy here. We expended energy into the environment to evaporate the water, and then expended energy to get it back out as water. Fancy that.

First and foremost, you need to focus on the fact that all energy generated by humans once fell on the Earth in the form of Sunlight and was employed in the process of photosynthesis. Everything we eat, animal or vegetable, came to us via photosynthesis. Essentially, all we as humans are doing is providing a mechanism to release this "banked" (kinetic, chemical) energy that has been delayed through the growth of plant life. Cows, pigs and chickens are no different than we are when it comes to relying upon plant life. The only significant difference is that they are concentrated in vast, agricultural regions with little or no population density.

I have previously covered the extent of our usage of, and reliance upon, fossil fuels. They are nothing but withdrawals from a bank of hydrocarbons generated over millions of years and stored in the underground vault from which we are extracting them. I am going to lump nuclear energy usage as it relates to the generation of electricity in with all other fossil fuel usage. No matter how we view this subject, all that usage ends up at some point as heat on, in or close to the surface of the Earth. *Table B-2*: **U.S. Fossil Fuel Contribution** illustrates through known facts, and some reasonable assumptions, the effect of the usage of fossil fuels within the geographic confines of the United States.

- **316.985** btu/ft$^2$/hour (**line 1.9**) is the maximum energy hitting the surface in a direct, normal fashion. In other words, perpendicular to the surface. See **"Parable Hour"** in **Part A**. Let's assume that this value is modified by angular deviation, and the fact that the Sun is not really "shining" 24/7. So, if we assume the best case, what we really get available on that square foot is the equivalent of 6 hours of direct, normal radiation. Therefore, the maximum daily value is **1901.911 net** btu/ft$^2$/day (**line 1.11**)[37].
- Now, most of the data that we have seen documented on solar collector panels by Electric Utilities demonstrates that they witness an overall "capacity factor" of about 25%, to the best-case 30%[38] [39]. This accounts for such variances as reflection, cloud cover, rain incidence, etc. Keeping that in mind, and assuming the most probable case of 25% (**line 1.12**), then the available energy that can be used to create weather is **475.48 net** btu/ft$^2$/day, or **0.1394 net** kWh/ft$^2$/day (**lines 1.13 & 1.14**).
- Given that we really are a pretty big country, we need to know how much energy is available over a larger area. There are **27,878,400** ft$^2$/mile$^2$ and **3,677,646** mile$^2$ in the United States [40] [**3,531,905** land and **145,741** inland waters and the Great Lakes, (**lines 1.1 & 1.2**)], including all 50 states and the District of Columbia. Doing the math, this translates into **0.003885 net** TWh/mile$^2$/day, or **13,721 net** TWh /day over land only (**line 1.16 & 1.17**). Given that there still are **365.25** days per year, then on the average we see **5,011,737 net** TWh/year (**line 1.18**) available to heat the land surface of the United States and provide the energy for photosynthesis and other uses.
- Stick with me, we're getting to the point. Take a look at *Table A-6*, referring to *Figure A-1* (**Part A**) which illustrates that the United States, all of them combined, consumed **27,240** TWh in 2020 (**line 1.21**). Remember, fundamentally, this all ends up in heat right at the surface of the Earth at some point in time. Essentially, this means that our consumption of fossil fuels is the equivalent of increasing the energy coming from the Sun by **0.54%** (**line 1.21**), which oh-by-the-way, is not something to sneeze at. Clouds don't get in the way of that degree of contribution.

  However, this is not the entire story. Remember that weather is local, influenced in the long term by factors and events that not only occur on-site, but "somewhere" else, e.g., the flight pattern of the butterfly in San Luis Obispo. Reference *Table B-3*: **U.S. Energy Re-Distribution**.

- The assumptions here are that the solar insolation and the total land area did not change from 1950 through 2020. However, per the referenced data, the total U. S. energy consumption increased by **183%** (**line 2.5**), from **10,140 TWh/year** to **28,658 TWh/year**; average 2016 through 2020 (**line 2.4**). The energy consumption both in the U. S. and world-wide decreased from 2019 simply due to the COVID-19 shutdowns and dramatic reduction in usage. This is why I am employing the 2016 to 2020 average in all energy usage data presented in this section as 2020.

- Utilizing classic sensitivity analysis assumptions, and stating the most severe case, let us postulate that around **10%** of the surface area of the U. S. is devoted to urban areas, meaning that only **10% of the total solar insolation** can affect the area. Likewise, under the same guise, I would propose that around **90%** of our consumptions of fossil fuels (**line 2.12**), and directly or indirectly via electricity usage, occurs in those same urban areas. Keep in mind, there are no "Amber Waves of Grain" or "Corn as High as an Elephant's Eye" in those urban areas, like Downtown Manhattan, to soak up the portion of the visible spectrum devoted to photosynthesis.
- Okay, given all of the above data, and assuming that all of the net solar insolence that falls on those areas is augmented by the human use of fossil fuel-derived heat, we come up with a startling result. **10%** of the U. S. solar energy is **501,174 net** TWh/year (**line 2.1 \* 0.1**). **90%** of our contribution from human-derived heat is **24,516** TWh/year (**line 2.12**). Lo and behold, our urban areas now look like microclimates of their own. This analysis just showed that on the average, the contribution to the heating of the urban areas from human usage of fossil fuels, when added to the solar contribution, easily can be shown to increase the energy added to the surface of the Earth from **0.1394 net** kWh/ft$^2$/day to **2.4898 net** kWh/ft$^2$/day (**line 2.13**). Taking this analysis one step further, to **80/20** (**80%** of the fossil fuel energy, **20%** of the land area), the result is **1.1066 net** kWh/ft$^2$/day (**line 2.9**). Neither one of these scenarios is a trivial amount of energy increase and needs to be examined further. Feel the burn yet? If not, you will shortly.

There is a final human factor that cannot be ignored, as you will witness. Though I have read in numerous publications that the contribution of humans is to be ignored, I decided to test this assumption, as any good engineer should in order to earn his/her pay.

- *Table B-4*: **U.S. & The World Population Influence** shows how this all plays out on the U. S. and World Stage. World population has increased from ~2.6 billion in 1950 to 7.8 billion in 2020[41] (**line 3.17**). The U. S. population likewise has increased from 3.9 million in 1790 to 150.7 million in 1950 to 331.4 million in 2020[42], per our Census Bureau (**line 3.17**). Taking a cue from the above data pertaining to the "Rule of Thumb" heat generated by the human body, this means that on the average each man, woman and child in the world generates 2.6 million btu/year sensible heat, and 1.8 million btu/year latent heat added to whatever else flows onto the surface of the Earth. This is a total of 4.4 million btu/year/capita, or 365 ton-hours of cooling, or heating, required to keep the human body in equilibrium with its surroundings each year. And you wonder why your Air Conditioning unit must work so hard each day.
  - ○ I must throw one of my standard tidbits of trivia into the mix/fray here. Probably, you have wondered endlessly, through nights of sleep deprivation, what the term "ton of air conditioning" means. Waste no more sleep, I'm here to save the day. It is derived from the fact that it takes 24 hours, at a rate of 12,000 btu/hr, to melt or freeze one ton, 2,000 lbs., of ice. It requires 144 btu/lb. of water to freeze it. Thus, doing the math, 2,000 \* 144 = 288,000 btu. When divided by 24, the answer is 12,000. Therefore, 1 ton of air conditioning equals a rate of 12,000 btu/hr.
- **Lines 3.1 to 3.3** are a repeat of data in *Table B-3*. **Lines 3.5 & 3.6** show the additional energy added to the surface of the earth within the U. S. along with the % of the overall solar insolation. The human "body" factor seems trivial until we continue the analysis into the urban areas as we did in *Table B-3*. **Lines 3.10 to 3.14** show that the human bodies resident in those urban areas increase the effect of the urban area's share of the overall solar insolation by 12% to 28%. From a micro-climate analysis standpoint, this obviously must be taken into account, or at least recognized.
- One other thing we need to keep in mind having to do with the energy expenditure illustrated in the "Fossil Fuel" column. I illustrated above that each human body requires ~12,000 btu per day (1 ton-day

of air conditioning) in order to remain in equilibrium with itself. This energy quantity must be multiplied several-fold per person when it is applied to our living and working environment. In other words, we spend a bunch of energy just to maintain that indoor environment just for our comfort.

- One other point that should be addressed in this section and that is the data presented in **Lines 3.16 to 3.20.** Effectively, this illustrates that in the World Community, though each body generates the same amount of energy, the energy contribution (as a % of TPES) (**Line 3.20**) of the aggregate of human population World-wide is substantially more than when compared to the U.S. alone. The answer is simple; we here in the U.S. have the luxury of consuming substantially more energy per capita than the rest of the World combined. Per the above, a lot of that overage is devoted to maintaining our physical comfort. OBTW, I have no problem with that observation.

The above urban-focused phenomena are real. I witness and document the fact that urban areas heat up substantially when compared to the "Country." Every day, 365.25 days per year, both my wife and I travel from a "Country" environment into an urban environment from midnight until 4:00 AM. "Country" is not densely populated suburbia. It's horses, cows and chickens "Country." The average ambient temperature difference, via my van thermometer is 5 - 10 °F, summer or winter. The absolute values are not material because my van thermometer has not been certified and calibrated by the National Standards Institute. The data is repeatable; it is the consistent differential that is the tell-tale.

# 4. The Earth's Surface, GHGs and the Weather

We have to assume that the data collected by NASA through the employment of the CERES project is accurate and that the average temperature of the Earth has increased about 1.2 °C, or 2.2 °F since the 1850-1900 baseline[43]. The CERES data gathering effort was a continuation of the Earth Radiation Budget Experiment (ERBE) which began in 1987. Theoretically, this data and the information it provides is unbiased and non-partisan. Also, the instrumentation up until the launching of the data gathering satellites was certainly not as robust and "state-of-the-art" as that which we currently have available. Most importantly, it is the temperature of the Earth's surface that drives the weather, so it's sure nice to believe that information is reliable. The real questions before the house are:

- For example, what was the weather like during the Civil War in 1861?
- What caused the change in the temperature of the surface of the Earth?
- How much of an impact is the use of fossil fuels as it relates to their concentration of use in highly populated, urban areas?
- Who is the bigger "enemy;" $CO_2$, Clouds and/or water vapor, energy flowing from the Earth's core, human interaction, . .? Is there really an "enemy?"
- From both a micro (urban area) standpoint or a macro (Pacific or Atlantic-spawned hurricanes) standpoint, how can we define if we are better off, or worse off, with the rise in global and/or local temperatures?

On with the show. Please take another look at *Figure B-1*. It shows not only the contribution of heat to the surface of the Earth, but the atmospheric temperatures as we proceed upward through the Boundary Layer (< 2,000 m) and into the Troposphere[44]. On the average, the temperature of the atmosphere decreases about 6.5 °C/km, or 11.7 °F per 3,300 ft as the altitude increases[45]. This decrease continues until we reach the Troposphere/Tropopause interface[46], around 11,000 km to 15,000 km, on the average, at which point the temperature remains constant at -56.5 °C, or -69.7 °F until reaching the Stratosphere. The Tropopause extends upward to 18,000 km to 20,000 km, on the average, where the Stratosphere starts. The Troposphere is where almost 100% of the weather action takes place. It contains over 90% of the GHGs and vast majority of the clouds. GHGs, other than water vapor, currently compose only 0.063% of all the atmosphere by mass, so why are we so worried about their buildup?

Clouds, and water vapor (good ole' $H_2O$) play a substantially larger role in our climate than all other GHGs combined. In fact, water vapor is the largest GHG that we must contend with. See *Table A-1*: **Earth's Atmosphere** in **Part A**. It's important to understand how and why water vapor is so important, but we will cover that issue more in **Section 7**. You will notice the temperature decreases as a function of altitude increase in **Figure B-2** which is a direct result of Adiabatic Expansion, to a degree, and atmospheric radiation losses/gains (see **Appendix A**), as opposed to any influence from convective or conductive heat transfer. As the air rises, it expands and cools simultaneously. This is a thermodynamic principle that is gospel and is fundamental to the operation of your household refrigerator and air conditioner.

The air is heated by the energy stored in/on the surface of the Earth, as given to it by the various sources mentioned above. The inherent turbulence in the Boundary Layer picks up this heat via forced convection, and conduction to a substantially lesser degree, and then rises toward the Tropopause where it stagnates. The tropopause is an inversion layer, much like the ones over Los Angeles and Denver, that traps air and heat under its mantle. The vast majority of the air in the Stratosphere and above is heated by absorption of the energy from the Sun, not from that stored in the Earth's Crust. {OBTW, there is very little thermal mass above the Tropopause.} Most importantly, not only does the cooler air that replaces the rising hot air pick up heat, but it also picks up moisture to one degree or another. Here be the key to trapping heat and giving us the source of our cloud layer(s).

Up to this point, most of the issues surrounding Earth's interaction with the rest of the Universe has been somewhat straight forward. We have dealt primarily with the input of energy into and onto spaceship Earth. However, we know from simple logic that in order to maintain an equilibrium of both temperature and our basic weather patterns, we cannot continue to accept all that solar energy without giving some of it back into space. In order to expel the excess energy, it must radiate from the surface of the Earth and the components of the atmosphere. It is very important to remember that we do not lose heat via convection or conduction because there is nothing in space to which heat energy can convect or conduct. In fact, even $H_2$ molecules, the lightest in existence, cannot easily escape Earth's gravitational pull. Space is a void, literally. In concept, this process of energy/heat loss is very simple. The energy contained in/on the surface will radiate toward a body or bodies that are at a lesser temperature. Classically, this is known as Black Body Radiation.

- **Heat Transfer via Conduction:** This is the simplest form of heat transfer because no portion of the substance has to move or change location to any appreciable degree. It can take place in both fluids and solids and can be described simply as an exchange between the molecules of a substance by them playing "bumper-car" and via intermolecular radiation. Molecule **A** vibrates due to heat content and "bumps" into molecule **B** which has less energy; the result of the collision is that **A** imparts a portion of its energy to **B**. As with radiation, this continues molecule by molecule until equilibrium is established, or more heat is added to the system at some point from outside of the system.
- **Heat Transfer via Convection:** This is a little more complicated because it involves only fluids, gas or liquid, coupled with the movement or displacement of the substance. Convection comes in two forms; **Natural** convection and **Forced** convection.
  - ❍ **Natural** convection occurs when a fluid is heated in part causing that part to expand and become less dense. This invokes a pressure difference, forcing the hotter, less dense portion of the fluid to rise due to gravity, and be replaced by a cooler portion of the fluid. Heat is then transferred to the cooler fluid and the cycle continues. An interesting side bar here is that within the International Space Station, where Gravity has no influence, Natural Convection can't occur. Give that as a quiz question to your neighbor at the next BBQ gathering in your backyard.
  - ❍ **Forced** convection essentially speeds up the process. The fluid is forced to move to, and then away from, the source of heat, at a speed faster than that imposed via Natural Convection, employing an external source of power applying the motive force. The net result is a faster transfer of heat at the expense of applying the energy necessary to move the fluid.
- **Heat Transfer via Radiation:** This is substantially more difficult to understand and quantify. At this point, you may want to visit & digest **Appendix A.** Classically, radiation occurs throughout the

entire electromagnetic spectrum, but what we are most concerned with is the ultraviolet, visible and infrared regions. This is how the energy from the Sun reaches Earth and imparts its energy via heat transfer or photosynthesis. The jury is out pertaining to the method of transfer, be it a "wave form," or a photon which could emulate "wave form" activity. Regardless, the energy transferred, at least in the regions we are most concerned with, is quantified using the Stefan-Boltzmann (S-B) equations below.

Radiation from the surface of the Earth, that which we are most concerned with, will take place in the infrared spectrum due to the temperature of the Earth. The equation quantifying the amount of energy transferred between two bodies of unequal temperature is known as the Stefan-Boltzmann Law[47][48]. Keep in mind that the absorptivity and emissivity of a body for radiation is the same quantity and is expressed in terms of %. A body with an emissivity ($\varepsilon$) of 90% will also absorb 90% of the radiation to which it is exposed.

## Stefan-Boltzmann Law

$P_{net} = P_{emit} - P_{absorb}$

$P_{emit}$ is equal to the value, in Watts or Btu/hour, of the radiant power emitted outward

$P_{absorb}$ is the amount of power a body absorbs from its surroundings.

For purposes of this document, we will assume that the emitting body (Earth's Surface) is always at a temperature greater than that of the absorbing, or receiving, body, e.g., outer space.

The overall equation for the net energy radiated from the Earth's Surface then becomes

$P_{net} = A * \sigma * \varepsilon * (T^4 - T_0^4)$

**A** is the area of the radiating surface; something of great importance, as you will see

$\varepsilon$ is the emissivity/absorptivity of the emitting surface

**T** is the temperature of the emitting surface

$T_0$ is the temperature of the absorbing body

$\sigma$ is the Stefan-Boltzmann Constant
$= 5.670374e-08$ W/m$_2$/°K$^4$ - SI Dimensions
$= 1.71400e-09$ btu/hour/ft$^2$/°R$^4$ - IP Dimensions

To put this in terms that we can all understand, let's examine what happens when we as humans occupy a conditioned space[49]. Because our body's survival temperature is ~98.6 °F, we will radiate heat from all our surface area in the infrared region of energy. In this range of temperature, both our body and our clothing have an emissivity and absorptivity very close to 1. Now, our clothing provides a degree of insulation, so if our surroundings are at a temperature less than our body temperature, then the surface temperature of our "envelope" is about 82 °F, ±. Assuming that the temperature of our surrounding, whether it be air, walls or your favorite lounge chair, is about 68 °F, then we will radiate energy, derived from our own metabolic processes, to those surroundings.  Any quantification of the energy that we radiate is proportional to the exposed area of our body, not the area of our surroundings, which are substantially larger.  The total of this energy turns out to be a heat loss value of ~100 W, or about 340 Btu/hour.  When looked at over 24 hours, this equates to about 2,000 kilocalories per day; the classic dimension defining what our food intact should not exceed.  Whadaya think about that?

Given that Earth's radiation is the subject, let's get back to the point. Looking at *Figure B-2*, you can observe three different depictions of radiant losses/gains via the size and direction of the arrows.  The large arrow is the net loss to outer space.  The next smaller arrow shows the effect of both absorption and reflection of the random cloud areas that show up periodically, basically unannounced. The little arrows represent the transfer to and from the nefarious GHG community.

We need to keep a perspective on this GHG issue.  As I have stated above, GHGs, aside from water vapor, only compose 0.0417 % of the atmosphere in terms of parts per million by volume, ppmv, or 0.0630% by mass, ppmm, as of the spring of 2021.  It is a molecule of $CO_2$, $CH_4$, $N_2O$, or $H_2O$, along with all other atmospheric gases, that both absorbs the radiant energy and re-emits that same amount of radiant energy. So, this means that our little ray, or photon, of radiant energy coming from the Earth's surface must bypass and otherwise avoid or encounter just under 1 million molecules of $N_2$, $O_2$, and the miscellaneous $H_2$, **Ar** or **He** atom in order to get at the 417 GHG molecules contained in that same amount of atmosphere. Obviously, that ray of infrared energy must be well aimed, because it has a 1 out of 2400 chance of hitting a GHG molecule.

## Side Bar: It's nothing but a game of Skeet:

As I mentioned, this radiant energy transfer stuff is really tough for most of us to get a handle on (See *Appendix A* for more discussion).  Most of us can grasp having a solid body, like the human body, give off and accept radiant energy.  We've experienced the absorption concept when standing in the sunlight, or before a steam radiator.  Likewise, the emittance issue is solved when we go out of doors on a cold winter morning.  The tough concept to grasp is radiation to and from a gas, especially one that radiates back to the emitter.

We will focus on the example of radiation from the surface of the Earth toward the sky dome.  Remember the Stefan-Boltzmann Law. **_Net_** radiant energy flows from hot to cold, not the other way around, as a function of the areas (**A**) of the emitter/absorber and their emittance ($\varepsilon$) values.  It's best to think of this form of radiation as photons rather than a wave form.  Visualize the surface of the Earth as a whole boatload of photon cannons, firing photon after photon micro-seconds apart.  Each photon carries an amount of energy from the surface and in the process cools it.  The goal of each photon is to reach outer space, thus causing its heat energy to be lost forever.

However, every once in a while, a photon gets real lucky and hits a molecule of GHG or water vapor, just like a skeet shooter periodically hitting a clay target.  The skeet shooter has the same problem the photon cannon has because the target is always moving.  Regardless, when that air-borne molecule is hit it gets "excited" and shoots back toward the surface an equally energetic photon, though not necessarily of the same energy value (jump ahead to *Figure AP-9*).  Again, see *Appendix A* for an in-depth discussion of the shooting back issue.  Essentially, conventional wisdom says, the "return fire" nullifies the original shot to a degree, thus not allowing the surface to cool as much as it otherwise would. You can now see how over time, conceptually, if there is a steadily increasing number of "targets" for the Earth-generated photons to hit, the surface of the Earth will cool less and thus increase in temperature. Of course this assumes all the energy imparted to that **GHG** molecule makes its way back to the Earth. Remember, the energy for our weather emanates from the surface of the Earth; ergo, the "conventional wisdom" mantra.

The term used for the loss of energy from the surface of the Earth through radiation is "Radiative Forcing" (RF)[50]. RF is expressed in terms of $W/m^2$ or $Btu/hour/ft^2$. The theory as postulated within the referenced article, hopefully using some _validated_ data, has postulated that with about 300 ppmv of $CO_2$, RF has a value of 260.12 $W/m^2$, or 82.5 $Btu/hour/ft^2$. The literature also says that an increase in density of $CO_2$ to 600 ppmv decreases RF to 256.72 $W/m^2$. The difference in energy transfer, $\Delta RF = 3.4\ W/m^2$, remains on Earth and provides heat that otherwise would not be there were it not for the additional atmospheric $CO_2$ content. Ergo, the published greenhouse effect. However, as you will see going forward when we apply the Stefan-Boltzmann equation, and OBTW some down-to-Earth logic and common sense, to the real-world data, we will get results that are substantially different than the above.

As a small note here, if $CO_2$ is the only metric used to gage what happens to the warming of the Earth, then whomever employs that strategy must be very cautious, outside of any validation of its science. When you view the results of the model I constructed, you will see demonstrated beyond doubt that the Earth's surface is warming due to the heat generated using all of our sources of energy, fossil or nuclear. It is extremely doubtful that any **GHG** has an appreciable affect. However, again if $CO_2$ is the only metric, and the concentration of $CO_2$ decreases over time as a function thereof, but the Earth still increases in temperature, the hue and cry will ratchet up a decibel or two. This trap would theoretically be sprung if, say, $CH_4$ replaced all coal but genertated the same level of heat. Look at **_Table C-1_** and you can see that the substitution of $CH_4$ for Coal would result in a 63% decrease of $CO_2$ emmissions, but have no effect on the increase in the temperature of the Earth's surface.

_Figure AP-9_ graphically represents what occurs when a **GHG** molecule is impacted with either a wave or a group of photons. Essentially, the molecule becomes "exited" when given energy and it immediately (like in picoseconds) dissipates that energy outward spherically. Many questions come to the forefront when viewing the figure. On a per molecule basis, how much of the aquired energy makes it's way back to Earth? What is the overall affect of having other **GHG**'s close by? Is this a "mutual admiration society" wherein they all keep the aquired energy to themselves and don't share, like we were supposed to do in grade school?

A phenominal amount of research has been conducted on radiant energy heat transfer within atmospheric environments containing $CO_2$, $H_2O$, $N_2O$, and other products of combustion of fossil fuels. This body of research data, documented in Marks' Mechanical Engineering Handbook[51], is used in the design and analysis of heat exchangers for furnaces and boilers. First of all, the radiative effect of $CO_2$ and $H_2O$ are not additive, even thought they are at the same temperature. Second, the driving factors regardless of temperature are the S-B constant ($\sigma$) and the 4th power of the absolute temperatures involved, ($T^4$). Third, the only variable of consequence is the product of the Area and the Emissivity, **(A * $\varepsilon$)**. As you will see in **_Appendix A_**, the critical issue is the product, **A * $\varepsilon$.** Most probably, it will shake out in the end that my projections give too much credit to the amount of energy making its way back to Earth.

Again, inquiring minds wonder if the dimensions given for Radiative Forcing factors quoted in the article mentioned are correct when viewed from other than the theoretical soapbox. Perhaps we are missing something, but the jury is still out. Because there is no "function-of-time" quoted, nor thermal masses involved, one must assume the power level of $W/m^2$ is in fact boundless, and Watthours play no role. We will explore this issue in depth in **Section 7**. Film at 2300 hours.

# 5. Let's Rewind to the Beginning & and do Some Analysis

Conventional wisdom, via the works of a bunch of smart Paleontologists, has it that the Earth started its journey about 4.54 billion years ago (BYA). There had to have been lots of horsing around with the matter allocated for the creation of Earth; you know, nuclear reactions, agglomeration of particles, gravity grouping, etc. Since that point in time, as you can imagine, a boatload of stuff has taken place. Look at *Figure B-4* and take a gander at what probably took place, temperature wise, since the "big bang."

Obviously, you can see that the Earth has gone through many "hot times" and "cold times," yet we are still here today. Note that our current period is "Icehouse" which began 33.9 million years ago, (MYA). Also obvious to the trained eye is that the next phase, beyond our control, OBTW, is a "Greenhouse" period. Fancy that! In fact, if we can believe what our eyes see, the cycles seem to get more active and the periods are shorter in duration. The real question before the house is that, given we have had no control over these cycles, what caused them?

- **Greenhouse Period:** A time when the temperature of the sea in the topics was ~82 °F and in the arctic it was ~32 °F. Essentially, there was no ice in the oceans or glaciers on land. As you can see in the diagram, the Earth has been in a greenhouse state for about 85% of its history.
- **Icehouse Period:** A time when the formation of ice in the form of polar caps and glaciers advances and recedes but doesn't disappear. These fluctuations are called "glacial" and "interglacial" periods. The cause of these fluctuations was discovered by a Serbian scientist named Milutin Milanković and have been subsequently termed as "Milankovitch Cycles." The cycles are caused by variations in the axial tilt of the Earth, from 21.5° to 24.5° every 41,000 years (23.4° is the "standard"), and the eccentricity of the Earth's orbit around the Sun. These kinds of variations cause substantial changes in the amount of solar energy reaching various portions of the Earth. You can imagine that this would have a great deal of influence on the seasonality of any location on Earth[52].

The term used for our past climate experiences is "paleoclimate." More than likely the main factors that precipitated the changes that have been surfaced is the confluence of fluctuations in atmospheric GHGs, most prominently being $H_2O$, $CO_2$ & $CH_4$, fluctuations in the Earth's orbit around the Sun, "wobble" or precession in the Earth's rotation, changes in the radiant output of the Sun, and random changes in the juxtaposition of the Earth's tectonic plates.

Studies have been done that firmly postulate that the $CO_2$ content of the atmosphere, during the Permian age (~299 MYA to ~252 MYA), fluctuated between the level at the turn of the century (1850 to 1900) of 250 ppmv, to upwards of 2,000 ppmv. Given that much $CO_2$ in the atmosphere, too bad FEMA wasn't around to give us all aid and comfort during our daily hurricane and tornado-driven disasters. That said, this large variance occurred during a period of Earth's evolution that contained massive growth in both plant and animal life.

When viewing *Figure B-5* from the "30,000 foot" level, several things pop out, at least to me. First, and probably foremost, we are looking at a span of time that is ginormous. 4.54 billion years is a really, really long time. But I guess if you just live it a day at a time, it ain't so bad. All funnin' aside, the Earth took a bunch of time to get its act together, stabilize its orbit, get the tilt right, and cool down enough to have its first icehouse period about 2.2 BYA. In the meantime, the fundamental elements of life (H, He, C, O, F, N, $CO_2$, $H_2O$) were formed which allowed the later development of photosynthesis and of course the byproduct $O_2$, the staple of aerobic life forms like air breathing animals.[53][54] The "Great Oxidation Event," which occurred ~ 2.33 BYA, was really the onset of the proliferation of life as we know it.

To me, one of the most interesting aspect of this figure is what occurred from about 600 MYA to the present. It was during this period that life began to flourish with abandon, die off and go through the process of being converted into the fossil fuels that we employ today to "fuel" our current, luxurious life style, and the Greenhouse / Icehouse cycles began in earnest. As I stated before, we are in the middle of the Late Cenozoic Ice Age, and I'm willing to bet, on our way to another Greenhouse period, the GHG issue aside.

If nothing else, *Figure B-5* should demonstrate to you that, most probably, something is happening to/ with/on/under the Earth that is moving us to be a little warmer; basically, completely beyond our control. I'm not smart enough or knowledgeable enough to know what that movement entails, but the engineer in me says that I ain't puttin' any money on the main driver being the buildup of GHGs in the atmosphere. In addition, I hesitate to even speculate that the short-term Earth warming is harmful. Given *Figures B-4 & B-5*, those folks that come after us, way after us, will have their hands full trying to keep cool. More later.

# 6. Let's Get to the Core of the Issue

At this point in time, I have heard almost nothing from the mainstream science community about one of the greatest assets we have here on Earth that is keeping us warm as opposed to freezing to death. Believe it or not, it is the backstop for the Sun's energy that even all the GHGs in the world could not overcome. It is the huge, almost unquantifiable in terms that we can understand, furnace at the core of Spaceship Earth. It is partly driven by numerous nuclear reactions that come in various forms, and some of the energy left over from the big bang during its formation. Regardless, if that furnace were to suddenly "go out," eventually the Earth would freeze over and never recover.

*Figure B-6* is a graphic representation of probably the best shot at the depth and temperatures of the Earth below our feet[55]. Spaceship Earth is a very complex amalgam of various compounds, temperature gradients, thermally caused mass movements, fluidics, etc. The core is basically composed of Iron-containing substances under enormous pressure and temperature. The lower Mantle is molten to some degree, moves about fluidly and is held in control mostly due to its high viscosity. It transfers its energy through forced convection and conduction, deriving most if not all of its energy from the core.

Because the Mantle is fluid, it is difficult to model from a heat transfer standpoint. The Upper Mantle and Lithosphere, however, are stationary and conduct their energy in a relatively stable, conductive manner. Thus, they are easily modelled as you will witness. The area that most affects our comfort, and to a degree our weather, is the Lithosphere which is composed primarily of silicon, aluminum, calcium and iron oxides. The ocean floor, for instance, is primarily calcium and magnesium carbonates that are the result of the depositing of "Ocean Creature" decomposition byproducts; animal, vegetable or otherwise.

*Table B-5* shows us many things about the crust of the Earth[56], coupled with some important characteristics, and principles, of some compounds that we see in everyday life. Given that weather, and our life here on Earth is totally driven by the transfer of heat, it is necessary for us to spend some time understanding what happens due to the influence on our surface temperatures derived from that furnace at our feet. Understand that the crust of the Earth is a necessary insulating layer that keeps us from frying to death.

Take the example of your kitchen oven. Turn it on to 350 °F to bake and when you come back 30 minutes later and touch the side. It's warm, but not to the point of burning you. As a side note, you would not want to have done that same thing with my paternal grandmother's oven. You didn't plug it into a wall outlet. It's source of heat was corncobs left over from the previous harvest and temperature regulated by gutfeel and experience in the kitchen. But I digress.

You will notice that all the compounds that compose the Earth's crust are oxides of one form or another. Remember in **Part A**, oxygen was formed early on, 4.5 +/- BYA and combined with anything close at hand, the first by-partisan element. The $O_2$ in our atmosphere didn't come along until the photosynthesis process began 3.5 +/- BYA. Thus, the crust was formed many moons ago and has shielded the surface ever since.

The columns worthy of note are the last three; "Thermal Density," "Thermal Conductivity," and "time travel." Thermal density is a way of getting a handle on the volumetric containment of energy. As you can see, the compounds which intuitively weigh the most are the least effective when viewed in the light of energy containment. The most effective, volumetrically, is water. Good old $H_2O$. See: **Parable Hour, Part A.** This, of course, happens to be why it is used in all our daily lives as a very efficient means of transferring heat using fluidics. Also, the most effective means of transferring heat using conduction is not water, but Silver, with Copper and Aluminum nipping at its heels, so to speak. Due to cost, obviously Silver is not the first choice to be used to manufacture the cooling coils in your residential air conditioner. Optimally, it is copper, but due to cost and manufacturability, the choice currently is aluminum.

The "time travel" column is important and I put it in the table in order to demonstrate one of the characteristics of any substance which is the time necessary to move heat from one location to another within a specific thermal mass. Note that the speed of movement through the crust is substantially faster than through the "dirt" at our feet.

Two more things we need to talk about before we move on to the next section. The first is pretty brief and involves the huge bank of stored energy in the Lithosphere. Just to give you something to Wow your next victim at the local neighborhood cookout, let me give you some facts to bandy about.

- The energy stored in only 1 °F of the Lithosphere is $7.94 * 10^{10}$ TWh, or $2.71 * 10^8$ Qbtu.
- At the yearly, energy consumption level of only the U.S. of 92.94 Qbtu, or 27,240 TWh, there is enough stored in that 1 °F to last 2,915,183 years. This is only the first 255 miles of the crust and only 1 °F out of thousands in the core.
- The total Solar Insolation that falls on the surface of the Earth each day is 3,056,264 TWh.
- In order to replenish that 1 °F of energy mentioned above, it would take 71.1 years of Solar Insolation alone, without any loss to Space, Photosynthesis, or any other chemical conversion process.

If nothing more, the above should demonstrate to you that whatever we do or don't do on the surface, doesn't even "scratch the surface" of the "problem," assuming we adopt, and hold dear to our heart, the view that weather change is a problem.

The next issue we need to dwell upon is the effect of the energy flowing from the core to the surface of the Earth, 24/7. Given the above data, it will take about 80 +/- days for a btu to make its way from the bottom of the Lithosphere to the bottom of our "dirt" layer. It will take another 8 +/- days to make it through that layer. All in all, we're looking at 90 to 100 days for any change, + or -, in temperature at the bottom of the Lithosphere to be "felt" on the surface.

Conventional wisdom says that we experience about 0.087 $W/m^2$ coming to the surface from the core, 24/7[57]. This equates to 2.09 $W/m^2$, 7.12 $btu/m^2$, 0.66 $btu/ft^2$ per day. This is equal to 0.14% of the solar energy at the surface. On the surface, no pun intended, this doesn't seem like much unless we don't, somehow or other, get rid of it. Only when it stays within the surface do we have a problem. Therefore, in any calculations pertaining to surface energy gain or loss, it must be considered. This energy is incremental to all other inputs, but an increment here, and an increment there, and pretty soon you have a lot of btu's to talk about. On a more sober note, implicitly, it is taken care of because if the surface temperature is relatively stable, it is being radiated to outer space. Not to worry!

# 7. Let's Talk About $CO_2$.

Please view *Table A-2* in *Part A*. GHGs are illustrated showing that water vapor, $H_2O$ is the greatest contributor, followed by carbon dioxide, $CO_2$, methane, $CH_4$, and nitrous oxide, $N_2O$. Ozone, $O_3$ is shown, but its effect is in the atmosphere above the Troposphere and has very little to do with any greenhouse effect on the surface. On the other hand, the other three gases are spread uniformly throughout the atmosphere in varying quantities, depending upon the distance from the Earth's surface due to the changing densities as we get farther and farther away from the surface (see *Figure AP-10*).

Numerous studies have been done that portray $CO_2$ as the evil one, though its effect as it pertains to weather fluctuations is quite questionable as far as I am concerned. My suspicion is that most studies have been funded by entities with an agenda, the most apparent of which has to do with the consumption and combustion of fossil fuels, probably in Western Europe. Given the obvious tiny quantities of both $CO_2$ and $CH_4$ in the atmosphere, I had difficulty believing that their effect is and would be catastrophic. Therefore, being the good, heat-transfer oriented engineer that I am, I decided to model the effects of the re-radiation of energy of the GHGs back to Earth.

As near as I can tell, most of the studies pertaining to GHGs in general focus on wave lengths, especially as they pertain to the energy from the Sun. This is not to say those studies are right or wrong because quite honestly, I have never studied light wave theory, especially in the region of the ultraviolet and visible. Primarily, I have always worried about radiation in the infrared region, which is the region that focuses on human comfort and the surface temperature of the Earth.

We talked about the Stefan-Boltzmann Law (**Section 4** above) and how it pertains to all radiation, most particularly from surfaces at or close to the average temperature of the Earth which we have stated as 59°F, ±10 °F, or 15°C. Obviously, the surface temperature of the Earth fluctuates substantially, but we will for purposes of this discussion focus on the 59°F & 15°C stated. Otherwise, we're going to start tripping over our shorts and get everyone confused, including me.

A key element for the equation is "**A**", the area that both radiates (emits) and absorbs. A surface cannot absorb quantities of heat more than the exposed area will allow, nor can it emit quantities of heat more than the exposed area will allow. As far as the Earth is concerned, the "radiator" of choice is the surface; huge and well defined, though molecular in nature. We must assume that all radiation from the Earth flows upward, and if stopped, does so by an absorbing surface of some defined area. The area question for the absorbing "surface" is not as easily defined or visualized, especially since we are talking about a gas.

Heat transfer, no matter whether convective, conductive or via radiation occurs due to molecular "excitement." In other words, a molecule regardless of what its containment medium is, transfers heat only because heat has been given to it and it gets "excited." This excitement will allow the transference either through physical contact or radiation and that molecule will become dormant until re-excited.

Okay, using this analogy, any GHG can only accept energy and re-radiate energy because it is "excitable." All the other molecular contents of the atmosphere cannot get very excited via radiation because their atomic bonds are so tight, the atoms literally can't vibrate appreciably in place. Don't get me wrong. All the other atmospheric gases accept and emit radiation. The issue is how much. Because their molecular numbers are so huge, it doesn't take a very large value of "$\varepsilon$" per molecule in order for the effect of, say $N_2$, to have some meaning in the overall heating of the atmosphere.

The Stefan-Boltzmann Law relies on area-to-area transfer. However, the GHG molecule, or any molecule for that matter, must be hit with some modicum of radiation in order to do its job. Again, go back to **Section 4** and re-read the part about skeet shooting. Given that the absorber(s) in this case is/are a gas, and moving at that, I decided to be as generous and conservative as possible, and assign the absorbing area a portion of the sky dome equivalent to the GHG molecular concentration by ppmv translated to ppm by mass, ppmm. The bottom line is that since the GHGs, other than water vapor, only comprise about 0.063% (see justified analysis below) of the volume translated to by-mass, that is the total absorbing area that I assigned to them. Meaning, only 0.063% of the "shots fired" by our radiation cannons could hit a GHG target and have it "shoot back." Bear in mind that $H_2O$ is the greatest contributor of all of the gases, but as you will see in *Appendix A*, its contribution fluctuates and can be assumed to be anything from 0% to almost 100%.

I used a piece-wise, nodal approach in the design of the model, assuming a unitized Earth surface area as the emitter, and as illustrated above, molecular sized receivers of each GHG, including nitrous oxide, $N_2O$ and $H_2O$. Each node of the model was the equivalent of the atmosphere broken into 1,000-foot elevations, up to 74,000 feet, which is well beyond the Tropopause and into the Stratosphere (see *Figure AP-10*). The model assumes that if a photon hits a GHG molecule it can go no farther and its loss from the surface is negated; again conservatism rules. Thus, any photon making it past those gases is assumed to take heat energy with it and thus cool the Earth's surface.

Going forward, we need to keep a couple of facts in mind. The assumption here is that we hold the Stefan-Boltzmann equation as the benchmark and that it rules when we calculate all the values pertaining to the transference of radiant energy. We need to accept that equation and its premise as scientific fact. Also, as the altitude increases, the atmospheric pressure decreases as does the temperature. These values are as presented by NOAA and other agencies including the CERES project.

1. The theoretical total amount of radiant energy that can possibly leave the Earth's surface, given the NOAA/CERES data assumptions that the average temperature of that surface is 59 °F and the temperature of Space is 4.86 °R, is 121.52 btu/hour/ft², or 383.1001 W/meter². This assumes that nothing is in the way to intercept any of the energy regardless of its form.
2. If this theoretical surface temperature is increased 2 °F, 59 °F to 61 °F, a value just shy of the 2.2 °F quoted above as the assumed temperature rise since the 1850 to 1900 period here on Earth, then the radiant loss is 123.32 btu/hour/ft², or 389.0434 W/meter².
3. This increase in the radiant loss, 1.88 btu/hour/ft², or 5.9433 W/meter², is important to remember as we go forward with this entire document. Remember this is "loss," not "gain." Therefore, to maintain equilibrium, we need to gain an equal amount else we freeze to death.
4. GHGs (Spring 2021 data), other than $H_2O$, compose 417.203 ppmv, 0.0417% by volume, throughout the atmosphere. Because the GHG molecules are larger than the other gas molecules in the air, I compared the molecular weight of the GHGs to the total molecular weight of the atmosphere. This increased the proportion of GHGs to 0.06299% by molecular weight. Given that the radiant

energy flow is from one "area" to another "area", I assumed that the area would be proportional to weight. Therefore, instead of having the GHGs comprise 0.0417% of the "area" looking upward, so to speak, I assumed the by-weight value of 0.063%. I felt this gave us a much more realistic and conservative answer.

5. Again, to be conservative, I assumed that when the GHG molecule was hit by that shot of radiation, it reradiated only in the downward fashion and all of its energy was used to negate the original "shot" content. In theory, the radiation should be spherical, but given the presence of the surrounding GHGs as receptors, I generously assumed the net finally transmitted back to Earth was as stated. Granted, it could be argued that the "100% reradiation" is both over-simplification and patently incorrect, but this assumption makes all results from the model very conservative. *Appendix A* expounds further.

6. The total weight of the atmosphere at standard conditions is 2116.224 lbs/ft². I took the model up to 74,000 feet. At that point, the remaining atmosphere weighs 93.86 lbs of air per ft². Therefore, the model looks at 95.56% of the atmosphere and by inference, 95.56% of the GHGs.

7. You will see in *Table B-6* a portion of the results of the analysis, which really is a very large, complex iterative mathematical model, using classic nodal relaxation techniques. *Table B-6* presents the amount of re-radiated energy from the atmosphere back to the Earth's surface, again using the above conservative assumtions, the results dimensioned as shown. I was able to modify the content of moisture in the air from 100% to 40%. Though the proportion of GHGs in the atmosphere are unaffected by altitude or temperature, this is not the case with the water vapor. Therefore, I had to take into account that whatever moisture was present at sea level could not survive the decrease in both temperature and pressure without condensing into aerosol form or clouds. Thus, when altitude and temperature was adjusted for each nodal elevation, the molecular percentage of water vapor had to be adjusted via standard, psychrometric principles as put forth in the ASHRAE Guide. I did not take into account the additional heat generated as a result of the condensation process, thus there is no bias in that respect.

8. The results show some substantial increases when $H_2O$ is adjusted, as in lines **4.1** and **4.3**. An adjustment of the $CO_2$ and $CH_4$ content, as in lines **4.1** and **4.5**, does not present a change as dramatic. The essence is that water vapor really plays a far more substantial role in the day-to-day variance of radiant forcing than that of the other GHGs. Note that any adjustment of $N_2O$ would not show anything significant, and thus no line items were presented.

Line item **4.9** is important because it is the baseline for our further demonstration of the results of the model. The concentrations of both $CO_2$ and $CH_4$ are pretty well documented by AGAGE and NASA[58] and the Mauna Loa data documented by Dr. Pieter Trans, NOAA/ESRL, and Dr. Ralph Keeling, Scrips Institution of Oceanography, *Figure B-9*. It is this data starting in 1950 and ending in 2020 that I used to map out the increase in concentrations of the two gasses in question. Again, I had to keep $H_2O$ and $N_2O$ constant in order to diminish the number of iterations that had to be used to get to some reasonable answers.

## Another Modeling Sidebar:

Modeling a complex heat containment/transfer system such as the Earth and its vicarious atmosphere is no trivial task. Almost by definition, it presents a classic "bean bag" problem. When you press in one side of the bean bag, it presents a hole or depression. If you only concentrate on the hole, the analysis is simple. However, there are a bunch of beans in the bag. If you don't look at the effect of the "poke" on all beans, you are losing sight of the overall effect of the poke. Most of the studies that I have read that focus on the increasing temperature of the Earth, concentrate only on basically one area; the increase of GHGs in the atmosphere as being the only cause to worry about. As you will see, this is far from the truth.

I have built a substantial number of mathematical models that focus on systems with complex heat transfer problems. One axiom that I have always relied upon is that "Energy in = Energy out." In other words, we must maintain equilibrium within the system as a whole, otherwise the results are in question. In line with that thinking, below are the assumptions that I made when designing the model, and the steps that I took each time the nodes were "relaxed." The nodes are "relaxed" when the system is in equilibrium for each "time-step" of the model and "Energy in = Energy out."

# 8. The Results You've All Been Waiting For.

I have employed classic heat transfer equations and known "constant" values used in any of the equations, as they are presented in the ASHRAE Guide and Handbook. Right now, they are too numerous to mention with one exception. I have assumed above all else that the quantification of any transfer of radiant energy is done via the dictums of the Stephan-Boltzmann Law, as illustrated above. As enumerated, the transfer is assumed to take place Area-to-Area. Also, the big driver for all radiation is Space at a temperature of 4.86 °R.

## Model Assumptions:

1. Re-radiation energy content is proportional to the atmospheric temperature of the GHGs as presented via the elevation values shown. In other words, the re-radiation values are less at the higher altitudes. Correspondingly, the densities of the atmosphere decreases with altitude as does the temperature. Therefore, when viewing both the temperature decrease and density decrease in conjunction with each other, the overall effect is reduced with an increase in altitude.

2. The land area in question is only the total of the United States as presented in *Table B-2*. Due to the difference in thermal densities of the land surface and that composed of water, they were separated via applied percentages as enumerated below.

3. The only fossil fuel consumption data used was that associated with the United States as a whole. As with weather, energy consumption and its contribution to the Earth's surface temperature is **Local**. The thermal energy associated with that consumption was assumed to be converted during the year in question and without any lag time involved due to any manufactured products that eventually ended up as "biomass" combustion.

4. Given that a substantial portion of the Sun's solar insolation is converted into vegetation that either decomposes (biomass) or is consumed by our population, no adjustment was made to compensate for lag time in this instance.

5. It is assumed that at the end of 1949, the United States was at equilibrium WRT Energy in = Energy out. In other words, if there was no adjustment in any of the three factors in question; GHG increases, fossil fuel consumption increases, and any increase in population, then the status quo would remain as 1949 to this day. It is only the increases from that benchmark that are being taken into account in the model.

6. No adjustment was made having to do with urban or population densities. The land, population and water were assumed to be equally distributed across the landscape, so to speak.

7. The thermal capacitance of the land was assumed to be 1,640.7 Qbtu/°F, comprising 96% of the total surface. The Thermal density was assumed to be 33.326 btu/ft³/°F, and the depth of the crust was assumed to be 500 feet. Even though the lithosphere is 1,346,400 feet deep, the effect of the core (as talked about above) below 500 feet should help maintain some sort of short-term stabilization. These two values, when multiplied by the land area of the United States, results in the capacitance. The value of this capacitance is an incredibly important flywheel in models of

this nature. This flywheel serves as the mathematical equivalent of the mechanical flywheel on an internal combustion engine's crankshaft. The size of the capacitance keeps the model stable and the order in which each of the nodes are mathematically "relaxed" is the equivalent of selecting the proper firing order for each cylinder in the engine. Also, it provides a mathematical indication of why thermally driven phenomena in our world happen so slowly. See the Urban-to-Country temperature difference that I mentioned above.

8. The thermal capacitance of the water volumes was assumed to be 38.0 Qbtu/°F, comprising 4% of the total surface. I used a thermal density of 62.32 btu/ft$^3$/°F and a depth of 150 feet. Though much of the Great Lakes and other glacier-made, water filled depressions are deeper than that, the constant convection currents and surface disturbances impose a blending mechanism that is hard to model. No coastal waters were considered as contributors.

9. The baseline temperature of the surface of the United States was assumed to be 59 °F, or 518.67 °R at the end of 1949. The receptor of Earth's radiation, outer space, was assumed to be 4.86 °R, or -454.81 °F and did not change throughout the model years. The Stefan-Boltzmann equation was used to determine the absolute best-case loss of energy from the surface of 121.5234 btu/hour/ft$^2$. Any increase in that value due to an increase in the Earth's surface temperature as the model moved forward "year-by-year" was subtracted from the total capacitance in order to relax that node.

10. Fossil fuel usage data is quite accurate, especially for the United States, from 1950 forward, as you have seen in **Part A**. The $CO_2$ and $CH_4$ data had to be extrapolated backwards (see *Figure B-9*), but it was quite linear in that region. The population data that was used came from the 10-year U.S. data as presented by the Census Bureau. Data for the intervening years was extrapolated assuming linearity.

11. The model only focuses on *differentials* from the end of 1949 to the beginning of the 1950 baseline forward. No absolute values will be presented. The energy balance goes like so:

    a. The base-line assumption for $CO_2$ was 310 ppmv and 1.42 ppmv for $CH_4$. The final values grew from there year-by-year, via the aforementioned data sources, to 415 ppmv $CO_2$ and 1.87 ppmv $CH_4$. $H_2O$ quantities and $N_2O$ quantities were assumed to remain at 40% relative humidity and .333 ppmv respectively.

    b. Any fossil fuel usage for the year in question is added to the thermal capacitance nodes. It is assumed that 98% of the fossil fuel usage accrues to the land node, and 2% accrues to the water node.

    c. Any population increase from the previous year, multiplied by the thermal values presented in **Section 3**, *Table B-1*, are added to the thermal capacitance nodes, 100% going to the land node.

    d. The contribution of Radiant Forcing is added to the capacitance nodes, with 96% going to the Land node and 4% to the water node. The assumption being that since we were considered to be steady-state at the end of 1949, and the ideal loss is as stated in "**9**" above, then any loss that was negated due to the greenhouse effect never left the surface and thus needed to be added back into the capacitance.

    e. The temperature calculated at the end of the previous relaxation is then used to determine a new, incremental loss from the thermal capacitances as compared to the ideal loss and is subsequently subtracted from the total of each capacitance. It is apportioned 96% to Land and 4% to the water.

    f. At the end of the node calculations, a new surface temperature is calculated for both the land and the water. These temperatures are then normalized via the surface percentages, 96% land, 4% water, to arrive at a new temperature for each node to be used at the start of the next nodal relaxation exercise.

g. All temperatures shown are °F, and all energy quantities are in btu or Qbtu.

h. No concentrations of thermal densities or capacitances due to Urban vs Country has been taken into account. Quite honestly, this granulation of the model would require a bank of old Cray computers with parallel processing capability that would be enormous. Frankly, this is why attempting to model with 100% accuracy, the weather going forward, or even backward, is a fool's errand and any results should be taken with a large cocktail, or maybe more.

We now get to the meat of this section, the results of the model and the interpretation of same.

*Figure B-7* graphically presents the accumulated United States surface temperature increases from 1950 through and including 2020. One thing is important to note. Given the way that models of this nature can be constructed, the absolute values shown are meaningful only when viewed in the light of the size of the thermal capacitances used. The bigger the capacitance, the lesser the values become. The balancing act that must be performed is to make sure the capacitances are not too small, but that the temperatures shown fall within the "reasonable" range. Smaller sizes mean greater absolute values, but it also adds to the instability of the model. The key is to understand the relative nature of each of the components. The absolute values literally can be "42'd" to anything the investigator wants, depending on the grantor, if the study is funded by a grant, or the political whims of a particular governmental agency. More on "42-ing" later.

The significance of the *Figure B-7* is simply represented in the relative size relationships. The **Green** area represents the net energy flowing into the thermal capacitance from the combustion of fossil fuels and the use of nuclear energy. The **Blue** area shows the effect of re-radiance back into the thermal capacitance from atmospheric, GHGs. The little **Red** dots show the minor significance of the additional energy added from our human existence through population growth. The obvious conclusion is that most of the temperature increase here within the United States is from our use of fossil fuels, not the incremental $CO_2$ that is a byproduct of the combustion of those same fossil fuels. Remember, the $CO_2$ contribution can be argued to be zero, thus the model results are conservative in nature.

It is important to note at this point the need to keep all nodes in the model "up to snuff," so to speak. In much of the dialogue that I have read pertaining to "Global Warming," seldom, if ever, is the effect of increased surface temperatures brought into the narrative. In fact, most models and explanations thereof concentrate on air temperatures rather than surface temperature changes resulting from banking and retrieving energy from large thermal capacitances.

Air, per se, has little thermal capacitance, and most importantly, doesn't stay in place long enough to capture any significant amount of energy. Whatever it acquires, it releases quite quickly. As a frame of reference, the thermal density of air is ~0.018 btu/ft³/°F within several hundred feet of ground level, as compared to 62.32 for water, and 33.33 for the Earth's crust. See "**Parable Hour**" **Part A.** Given this information, you can see why a good thermal model design, when used to track huge quantities of energy, simply ignores the capacitance effect of air. It is a great conduit to transport heat quickly using forced convection, from one location to another, but other than that, it cannot really affect our model.

*Figure B-8* graphically illustrates the effect of re-radiation to space when the temperature of the emitter, Earth's surface, has increased in temperature above the baseline. As you can see, the overall effect at the completion of this 71-year model is that 0.2 °F were lost to space due to re-radiation. On the surface, this may seem trivial, and it is if viewed only in the light of the entire United States surface area as the emitter. As mentioned above, large urban areas see surface temperatures at night far in excess of those recorded in the vast, agricultural regions. As a reminder, the amount of energy transferred to space through radiation is a fourth-order function of the temperature difference between the emitter and the absorber. Time for another sidebar.

> **The Urban Side Bar**. Let's assume that we had an urban community with an area of 10 square miles. Also, let's assume that the surrounding countryside had an average, daily surface temperature of 59 °F (our old favorite), and that the urban area had a temperature 10 °F higher, or 69 °F. The energy lost through radiation from the same 10 square miles of corn fields would be 812.5 billion btu/day, or 9,922,248 kWh/day. However, the urban community would lose 877.0 billion btu/day, or 10,709,871 kWh/day. This is a net loss of 787,623 kWh (7.9% increase) per day, or enough to provide 50 average Florida residences electricity for a year, assuming it was electrical power and not heat. Don't take your eye off the ball; it's the demonstration of the difference that is important, not the absolute values of each temperature.

I have no doubt that you are pondering, right now, why I bring this subject to the front burner. Well, let me tell you why, just so that you can sleep better at night. Remember the issue that I brought up, both U.S. oriented and world-wide having to do with population density and the areas associated with those concentrations. It is highly possible that any global loss in radiation, and the associated perception of increases in surface temperatures, has been affected by the huge thermal concentrations illustrated, and thus any radiation losses are really augmented substantially by convective losses from the massive air columns, acting in a chimney fashion, rising above the urban areas. Keep in mind that I said air is a horrible bank for energy, but a great transmitter through convection. In this case, the convection is natural, but powerful just the same.

So, this brings to the forefront the obvious question. Are we really getting the right (read: accurate) information from the "experts," or are they off in some distant laboratory, contemplating their navel, and satisfying the grantor or educational institution with results that the funding entities want to hear? Inquiring minds wonder if we are getting our money's worth from the U.N., at least on this subject

# 9. Who Profits from Ringing the "Climate Change" Bell?

Again, we have inquiring minds at work; perhaps even over-time. On the surface, the Green New Deal is focused on the reduction of $CO_2$ and other GHGs in the atmosphere, and to a lesser extent, some cosmetic, but realistic and necessary, environmental concerns. Keep Europe in mind, given the huge quantities of fossil fuels that it must import. Air and water pollution come to mind, but I have difficulty drawing a straight line between "saving the trees" and electric vehicles. Reducing livestock flatulence seems a little Quixotic at best. Reducing the need for "hard copies" seems noble if only to keep trees growing so they can suck $CO_2$ from the atmosphere.

Is it the petroleum industry that is "fueling" this endeavor? Likewise, the coal industry? On the surface, why would they want to limit, and ultimately substantially reduce, their revenue by promoting any reason to supplant fossil fuels with any renewable energy source? Given the greedy, capitalistic nature of those land or lease-hold owners whose sole purpose in life is to destroy nature in the raw, one would think this an anathema.

Hold the presses, there could be something else to consider. Let's pretend (I love "let's pretend" games) you are a fossil fuel guru and mogul and you know you are sitting on, and/or own, a resource that is finite, and the market for that resource will never go away. It's like money in the bank. In fact, it's the only money in the bank and you have total control over when and how to spend it. Your thinking goes further down that road and you say to yourself, "self," why don't I let a bunch of "tree-huggers" force me to reduce production because electric cars become the vogue and I have to leave all those fossil fuels in the ground. All I have to do is maintain business in a cash-flow breakeven manner, garner my resources through proper stewardship, and wait for the time when I can recoup all my "losses" by letting the free market (the supply and demand function) drive the price up. Probably, a competent financial forecaster can make a good case using that logic.

Maybe we need to look to the nuclear industry. They have been needlessly put out to pasture, so to speak. However, the need for nuclear base load plants to supplant the aging coal plants that produce a bunch of $CO_2$, and smaller peaking plants the size of those used in nuclear submarines, will blossom soon enough. Renewable energy production is a great concept, but it has some systemic problems that I put forth in **Part A** which cannot be ignored. We will still need some bodacious sources of electricity production that really can't be satisfied with all the solar panels and windmills in the world (read: Texas, 2020).

Whoa! I forgot about the manufacturers of devices/hardware that will be used to harvest the energy from the Sun and convert it to electricity. There is no doubt that they will benefit greatly from the movement away from the use of fossil fuels to generate electricity. We couple that with the group (maybe one-in-the-same) who build the equipment to recharge electric vehicles. Or how about all the HVAC manufacturers who will have to phase-out the production of fossil-fired furnaces/boilers and begin manufacturing (and selling and installing) electric replacements. Or the manufacturers of electric transmission and distribution equipment, all the way down to the residential level.  There is a boatload of money to be made retrofitting all those commercial buildings and residential structures.

And I wouldn't discount the automobile manufacturers too quickly.  Granted, they are going to be building transportation vehicles regardless. However, at least currently, electric vehicles come with a big price tag, and perhaps a bigger profit margin.  I don't know because I'm not in that business.

On the other hand, as with any movement du jure, we could have a bunch of Crusader Rabbits running around, propagating, and not one of them really knows what they are waving the flags about.  They simply are waving flags to be waving flags, getting a lot of attention, and making the rest of our lives miserable. *c'est la vie*.  They happen to be getting a lot of tax-deductible donations, some of which in the manner of the BLM movement, could be buying those future vacation homes.

In any event, to find the real answer, we have to follow the money.  In our world today, nothing happens until something is sold, and there always is a price on the "something."

Film at 2300 hours . . . or not.

# 10. Have we been "42'd" (*)?

I think the number 42 is something we should discuss because it points to a larger problem that most of the citizens of the world face today. First, the "legend."

## The Douglas Adams Sidebar

The number 42 first came to my attention when I tried to read Douglas Adams' "The Hitchhikers Guide to the Galaxy." I never completed the book because it was so disjointed and esoteric, it was tough to hold my attention. Regardless, in the book, the protagonists Lunkwill and Fook, aliens who were also programmers, built a super-computer named "**Deep Thought**." They asked of **Deep Thought** the obvious question; what is the answer to the Ultimate Question of Life, the Universe and Everything, or words to that effect.

**Deep Thought** worked on the answer for 7.5 million years and finally told the ancestors of Lunkwill and Fook, Loonquawl and Phouchg, the answer. It was "42." Much speculation abounded at that point about what "42" meant and why it was so significant. Douglas Adams later in an interview simply stated it was a small number that he picked at random because it sounded good.

The legend, however, is much more glamorous and frankly I would much rather believe it than the "picked-at-random" explanation. "42" in the parlance of decimal, ASCII code is the asterisk "*". The asterisk in hex, it is 002A, and (nibble) binary 0010 1010, but it's the decimal version that is of import. When we employ a search engine to find something that we only know part of, we substitute the "*" for the unknown section of the search string. In other words, the engine will come back with whatever it wants to substitute for the "*" in order to fill in the blanks at the bottom of the page.

Thus, as the legend goes, when Deep Thought answered "42," what he/she meant was that the answer to the question could be anything you want it to be. I don't know about you, but I like that better than the random "picked-out-of-the-air" answer Douglas Adams voiced.

Or maybe we have fallen victim to another "Zohnerism." This is a term coined as the result of the outcome of a 1997 science fair project authored by a 14-year-old student named Nathan Zohner. He leveraged a concept-turned-parody founded in the 1980's by various groups & major media publications to see how gullible the scientifically vacant public could be. Water, good old $H_2O$, was called dihydrogen monoxide and the thrust of the science fair project was to get votes from 9th grade students as to whether or not dihydrogen monoxide should be banned. It turns out that 43 out of 50 students voted to ban it. Do you suppose that the scientists whose reports are being bally hoed by the IPCC as truth might be participating in a science project run amuck? Inquiring minds, again, wonder.

Okay, how does that apply to this discussion? I have felt from the beginning what I have stated before in this treatise; the $CO_2$ hue and cry is a lot of fluff with no substance. I've spent a lot of real estate here, and trees, I might add, to tell you why this probably is the case. Therefore, the only conclusion that I can rationally reach is that the increase in GHGs can mean anything you want it to mean. As Marx (I think), Hitler, or Goebbels (maybe) when acting as the chief propagandist for the Nazi Party, rightfully said that if you tell a lie loud enough and often enough, it becomes truth. What I have pointed out above sorta points in that direction. Many entities, rightfully or wrongly, want the $CO_2$ controversy to be what they want it to be.

I rest my case.

# 11. What is Life-After- EV Going to be Like?

One huge advantage of moving much of the energy consumption from fossil fuels to renewable sources is a reduction in the thermal energy build up in the surface of the Earth.  See **Part C** for the full scoop.

# 12. We Come to the End . . Finally.

Our problem as a society today is simple, at least here in the United States.  We have a huge educational void when it comes to the practice of rational thinking and questioning statements or dictums by searching for truth.  The skills associated with the practice of logic as applied to both life and any science or mathematical endeavor are not being taught either at home or in our school system.  Teachers and educators are more interested in telling us how we feel, or how we should worry about our skin color or that of our neighbors.  Quite honestly, this trend must be reversed, or our society will degrade just as all other great societal experiments have over the centuries.

First, we need to locate means to continually educate the populous pertaining to the facts, a word which has the WOKE definition of "the media's opinion."  Unfortunately, what we need to impart to all inquiries in the classic search engines, would be deemed as "disinformation," using the current filtering technology employed by social media giants.  If I attempted to publish any part or portion of what I have put forth in both **Part A** and **Part B**, it would be taken down because the fact checkers could not "verify" the validity of my arguments.  There should be no surprise here, because what I am saying is the equivalent of heresy in the social and mainstream media world.  This document does not fit the current (2021) narrative of either the social, print or televised media.

We must start at the kindergarten level and begin teaching students as we were doing in the early to mid-20th century.  Readin', writin' and 'rithmatic, coupled with some real science (as opposed to pseudo-science) background, and we may be able to recover.  Kids must be taught to question conventional wisdom and discover the truth on their own.  The answer is school choice in order to leave those educational decisions up to the parents.  We need to get the Federal Government out of the classroom and let the states and localities, who know the kids best, provide the curriculum.  The Teacher's Unions aren't helping us either.  For whatever reason, they have forgotten that their job is to educate by teaching truth, and not to be re-writing history to suit their current social agenda.

Let's move on to **Part C** and find out what the IPCC should really be talking about, rather than what they are talking about.

# Tables and Figures Part B

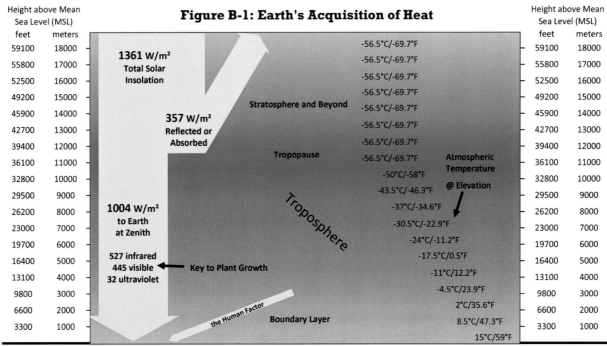

## Figure B-1: Earth's Acquisition of Heat

| Height above Mean Sea Level (MSL) | |
|---|---|
| feet | meters |
| 59100 | 18000 |
| 55800 | 17000 |
| 52500 | 16000 |
| 49200 | 15000 |
| 45900 | 14000 |
| 42700 | 13000 |
| 39400 | 12000 |
| 36100 | 11000 |
| 32800 | 10000 |
| 29500 | 9000 |
| 26200 | 8000 |
| 23000 | 7000 |
| 19700 | 6000 |
| 16400 | 5000 |
| 13100 | 4000 |
| 9800 | 3000 |
| 6600 | 2000 |
| 3300 | 1000 |

1361 W/m² Total Solar Insolation

357 W/m² Reflected or Absorbed

Stratosphere and Beyond

Tropopause

1004 W/m² to Earth at Zenith

527 infrared
445 visible
32 ultraviolet

Key to Plant Growth

Troposphere

the Human Factor

Boundary Layer

-56.5°C/-69.7°F
-56.5°C/-69.7°F
-56.5°C/-69.7°F
-56.5°C/-69.7°F
-56.5°C/-69.7°F
-56.5°C/-69.7°F
-56.5°C/-69.7°F
-56.5°C/-69.7°F
-50°C/-58°F
-43.5°C/-46.3°F
-37°C/-34.6°F
-30.5°C/-22.9°F
-24°C/-11.2°F
-17.5°C/0.5°F
-11°C/12.2°F
-4.5°C/23.9°F
2°C/35.6°F
8.5°C/47.3°F
15°C/59°F

Atmospheric Temperature @ Elevation

Mean Sea Level (MSL)

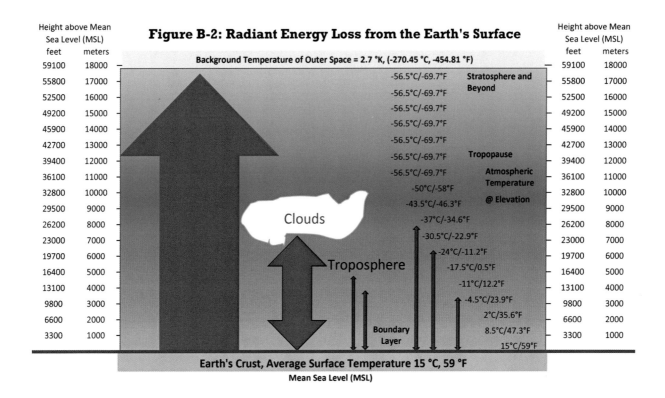

## Figure B-2: Radiant Energy Loss from the Earth's Surface

**Background Temperature of Outer Space = 2.7 °K, (-270.45 °C, -454.81 °F)**

| Height above Mean Sea Level (MSL) | |
|---|---|
| feet | meters |
| 59100 | 18000 |
| 55800 | 17000 |
| 52500 | 16000 |
| 49200 | 15000 |
| 45900 | 14000 |
| 42700 | 13000 |
| 39400 | 12000 |
| 36100 | 11000 |
| 32800 | 10000 |
| 29500 | 9000 |
| 26200 | 8000 |
| 23000 | 7000 |
| 19700 | 6000 |
| 16400 | 5000 |
| 13100 | 4000 |
| 9800 | 3000 |
| 6600 | 2000 |
| 3300 | 1000 |

-56.5°C/-69.7°F — Stratosphere and Beyond
-56.5°C/-69.7°F
-56.5°C/-69.7°F
-56.5°C/-69.7°F
-56.5°C/-69.7°F
-56.5°C/-69.7°F — Tropopause
-56.5°C/-69.7°F — Atmospheric Temperature @ Elevation
-50°C/-58°F
-43.5°C/-46.3°F
-37°C/-34.6°F
-30.5°C/-22.9°F
-24°C/-11.2°F
-17.5°C/0.5°F
-11°C/12.2°F
-4.5°C/23.9°F
2°C/35.6°F
8.5°C/47.3°F
15°C/59°F

Clouds

Troposphere

Boundary Layer

**Earth's Crust, Average Surface Temperature 15 °C, 59 °F**

Mean Sea Level (MSL)

## Table B-1
### Human Energy Contribution

| Activity | Sensible btu/h | Latent btu/h |
|---|---|---|
| "Rule-of-Thumb" | 300 | 200 |
| Seated, light work | 245 | 155 |
| Light bench work | 275 | 475 |
| Walking, light machining | 375 | 625 |
| Gymnasium | 710 | 1090 |

## Table B-2: U.S. Fossil Fuel Contribution

| | | | | line |
|---|---|---|---|---|
| [2]Area of United States | 3,531,905 | Land mile² | | 1.1 |
| Includes only Inland water and the Great Lakes | 145,741 | H₂O mile² | | 1.2 |
| | 3,677,646 | Total mile² | | 1.3 |
| | 9,158,022 | Land Km² | | 1.4 |
| | 377,768 | Water Km² | | 1.5 |
| | 9,535,790 | Total Km² | | 1.6 |
| | 2.79E+07 | ft²/mile² | | 1.7 |
| Total U. S. Land Sq Ft | 9.85E+13 | ft² | | 1.8 |
| insolation, btu/ft²/hour | 316.985 | | | 1.9 |
| hours/day | 6 | | | 1.10 |
| Total Insolation per day | 1901.911 | | | 1.11 |
| Capacity Factor | 25% | | | 1.12 |
| Net btu/ft²/day | 475.48 | | | 1.13 |
| Net kWh/ft²/day | 0.1394 | | | 1.14 |
| Net kWh/mile²/day | 3,884,982 | | | 1.15 |
| Net TWh/mile²/day | 0.003885 | | | 1.16 |
| U.S. Net TWh/day | 13,721 | | | 1.17 |
| U.S. Net TWh/year solar insolation | 5,011,737 | | | 1.18 |
| U.S. Net TWh/year Fossil fuel consumption | 21,377 | 0.427% | | 1.19 |
| U.S. Net TWh/year Other electric Generation | 5,814 | 0.116% | | 1.20 |
| | 27,240 | 0.544% | | 1.21 |

[2]www.census.gov/geographies/reference-files/2010/geo/state-area.html

# Table B-3:  U.S. Energy Re-Distribution

|  | 1950 | 2020 | Avg '16 to '20 | line |
|---|---|---|---|---|
| U.S. Net TWh/year solar insolation | 5,011,737 | 5,011,737 | 5,011,737 | **2.1** |
| Total US Sq Ft | 9.85E+13 | 9.85E+13 | 9.85E+13 | **2.2** |
| Net kWh/ft²/day | 0.1394 | 0.1394 | 0.1394 | **2.3** |
| U.S. Net TWh/year Energy Usage | 10,140 | 27,240 | 28,658 | **2.4** |
| % increase from 1950 | 0.00% | 168.63% | 182.61% | **2.5** |
| Uniform % of Solar Insolation | 0.202% | 0.544% | 0.572% | **2.6** |
| Densely Populated Areas = 20.0% of Total U.S. Land Area | 1.97E+13 | 1.97E+13 | 1.97E+13 | **2.7** |
| Densely Populated Areas consume 80.0% of Total U.S. Energy, in TWh | 8,112 | 21,792 | 22,926 | **2.8** |
| Net kWh/ft²/day in Urban Areas | 0.4119 | 1.1066 | 1.1642 | **2.9** |
| Differentiated % of Solar Insolation | 296% | 794% | 835% | **2.10** |
| Densely Populated Areas = 10.0% of Total U.S. Land Area | 9.85E+12 | 9.85E+12 | 9.85E+12 | **2.11** |
| Densely Populated Areas consume 90.0% of Total U.S. Energy, in TWh | 9,126 | 24,516 | 25,792 | **2.12** |
| Net kWh/ft²/day in Urban Areas | 0.9269 | 2.4898 | 2.6195 | **2.13** |
| Differentiated % of Solar Insolation | 665% | 1787% | 1880% | **2.14** |

# Table B-4: U.S. & The World Population Influence

| | 1950 | | Avg '16 to '20 | | Line |
|---|---|---|---|---|---|
| U.S. Net TWh/year solar insolation | 5,011,737 | | 5,011,737 | | **3.1** |
| Total US Sq Ft | 9.85E+13 | | 9.85E+13 | | **3.2** |
| Net kWh/ft²/day | 0.13935 | | 0.13935 | | **3.3** |
| | Fossil Fuel | Human | Fossil Fuel | Human | **3.4** |
| U.S. Net TWh/year Energy Usage | 10,140 | 193.5 | 28,658 | 425.5 | **3.5** |
| Uniform % of Solar Insolation | 0.202% | 0.004% | 0.572% | 0.008% | **3.6** |
| Densely Populated Areas = 20.0% of Total U.S. Land Area | 1.97E+13 | | 1.97E+13 | | **3.7** |
| Densely Populated Areas consume 80.0% of Total U.S. Energy, in TWh | 8,112 | 154.76 | 22,926 | 340.39 | **3.8** |
| Net kWh/ft²/day in Urban Areas | 0.4119 | 7.86E-03 | 1.1642 | 1.73E-02 | **3.9** |
| Differentiated % of Solar Insolation | 296% | 5.639% | 835% | 12.403% | **3.10** |
| Densely Populated Areas = 10.0% of Total U.S. Land Area | 9.85E+12 | | 9.85E+12 | | **3.11** |
| Densely Populated Areas consume 90.0% of Total U.S. Energy, in TWh | 9,126 | 174.11 | 25,792 | 382.93 | **3.12** |
| Net kWh/ft²/day in Urban Areas | 0.9269 | 1.77E-02 | 2.6195 | 3.89E-02 | **3.13** |
| Differentiated % of Solar Insolation | 665% | 12.689% | 1880% | 27.908% | **3.14** |
| | United States | | World | | **3.15** |
| 2020 TPES, TWh | 28,658 | | 166,979 | | **3.16** |
| 2020 Population | 331,449,281 | | 7,795,000,000 | | **3.17** |
| kWh per year per capita | 86,463 | | 21,421 | | **3.18** |
| Human Energy Expelled, TWh | 425.5 | | 10,006 | | **3.19** |
| Human % of TPES | 1.48% | | 5.99% | | **3.20** |

# Figure B-4: Earth's Temperature History

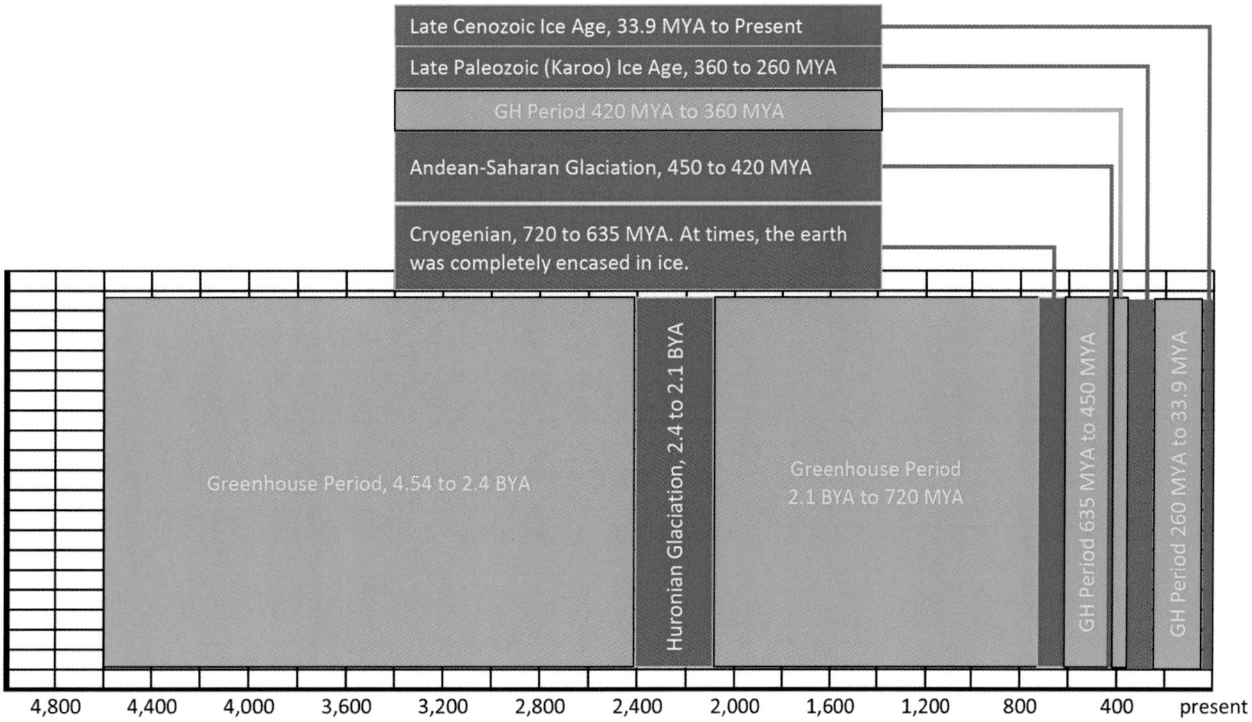

Late Cenozoic Ice Age, 33.9 MYA to Present

Late Paleozoic (Karoo) Ice Age, 360 to 260 MYA

GH Period 420 MYA to 360 MYA

Andean-Saharan Glaciation, 450 to 420 MYA

Cryogenian, 720 to 635 MYA. At times, the earth was completely encased in ice.

Greenhouse Period, 4.54 to 2.4 BYA

Huronian Glaciation, 2.4 to 2.1 BYA

Greenhouse Period 2.1 BYA to 720 MYA

GH Period 635 MYA to 450 MYA

GH Period 260 MYA to 33.9 MYA

| 4,800 | 4,400 | 4,000 | 3,600 | 3,200 | 2,800 | 2,400 | 2,000 | 1,600 | 1,200 | 800 | 400 | present |

[9]Greenhouse - Icehouse periods . . 4600 MYA to Present

[9]https://en.wikipedia.org/wiki/Greenhouse_and_icehouse_Earth#Greenhouse_Earth

## Figure B-5: The Development of Life on Earth

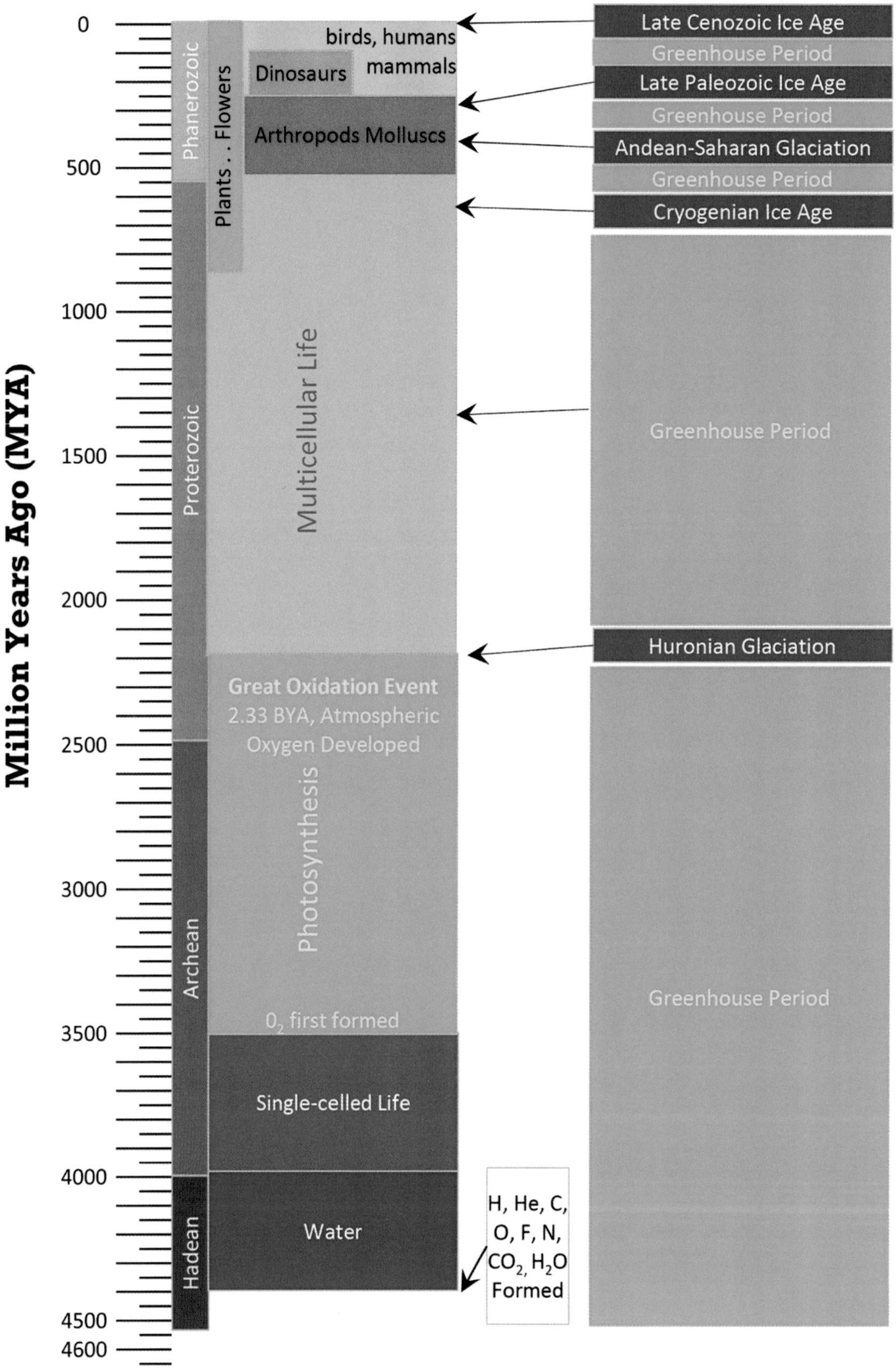

# Figure B-6: Earth's Temperature Gradient . . Core to Crust

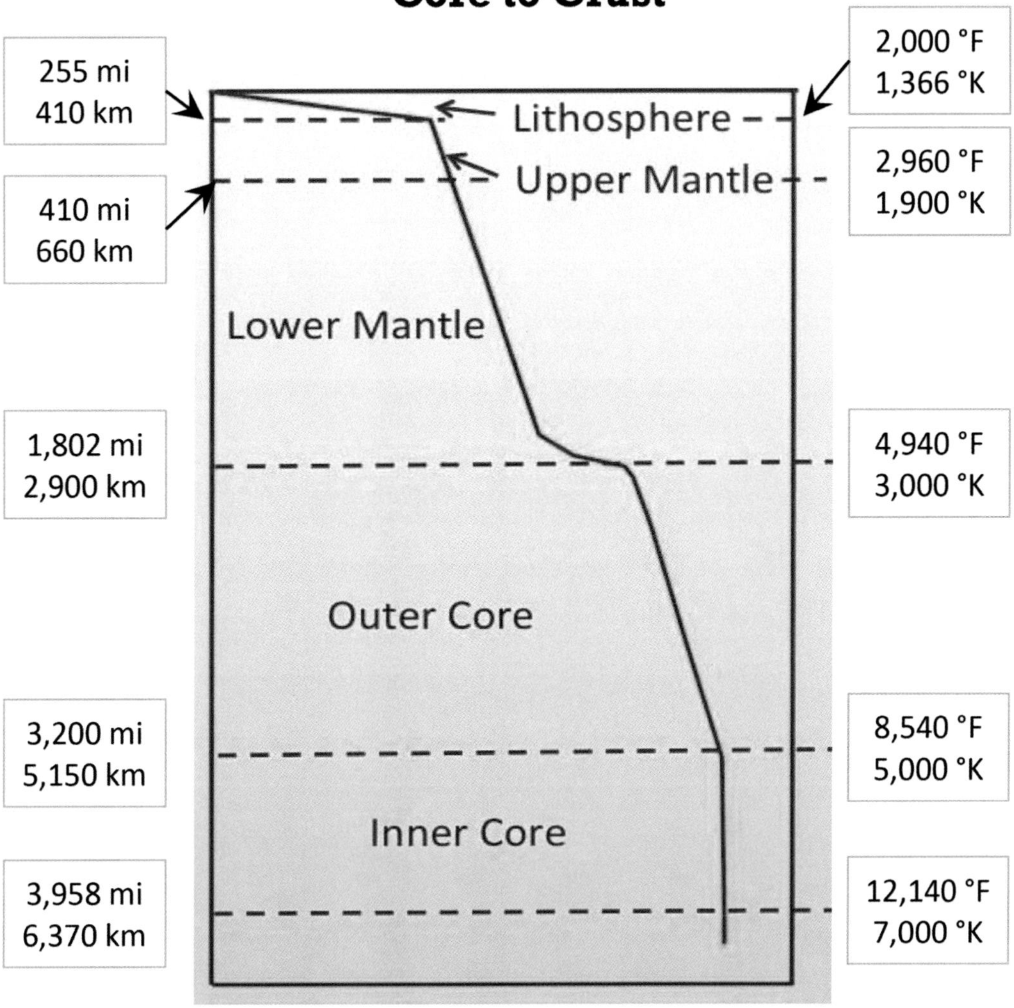

| | |
|---|---|
| 255 mi / 410 km | 2,000 °F / 1,366 °K |
| | Lithosphere |
| | Upper Mantle — 2,960 °F / 1,900 °K |
| 410 mi / 660 km | |
| Lower Mantle | |
| 1,802 mi / 2,900 km | 4,940 °F / 3,000 °K |
| Outer Core | |
| 3,200 mi / 5,150 km | 8,540 °F / 5,000 °K |
| Inner Core | |
| 3,958 mi / 6,370 km | 12,140 °F / 7,000 °K |

# Table B-5: Some Characteristics of Earth's Crust

| | Compound | Formula | Composition Continental | Composition Oceanic | Thermal Density, btu/ft³/°F | Thermal Conductivity btu/hr/ft/°F | [A]travel time 1 btu 50 ft |
|---|---|---|---|---|---|---|---|
| [B] Earth's Crust | Silica | $SiO_2$ | 60.60% | 50.10% | Continental = 40.3 | Continental = 3.31 | 22 min |
| | Alumina (Clay) | $Al_2O_3$ | 15.90% | 15.70% | | | |
| | Lime (limestone) | CaO | 6.41% | 11.80% | | | |
| | Magnesia | MgO | 4.66% | 10.30% | | | |
| | Iron Oxide | $Fe_2O_3$ | 6.71% | 8.30% | | | |
| | Sodium Oxide | $Na_2O$ | 3.07% | 2.21% | Oceanic = 38.1 | Oceanic = 3.24 | 22 min |
| | Potassium Oxide | $K_2O$ | 1.81% | 0.11% | | | |
| | Titanium Oxide | $TiO_2$ | 0.72% | 1.10% | | | |
| | Phosphorus Pentoxide | $P_2O_5$ | 0.13% | 0.10% | | | |
| | Manganese Oxide | MnO | 0.10% | 0.11% | | | |
| Soil Layer 300 ft | Dry | Clay - Water - Sand Mixture | | | 20.90 | 0.037 | 32 hours |
| | Saturated | | | | 33.33 | 0.1303 | 9.1 hours |
| Common Substances | Water | $H_2O$ | | | 62.3 | 0.348 | 3.4 hours |
| | Steel | Alloy of Fe, C, Ni, etc. | | | 58.7 | 26.2 | 2.7 min |
| | Iron | Fe | | | 58.2 | 34.9 | 2.0 min |
| | Aluminum | Al | | | 36.6 | 128 | 33 sec |
| | Copper | Cu | | | 51.2 | 227 | 19 sec |
| | Silver | Ag | | | 36.6 | 245 | 17 sec |

{A} This analysis assumes a solid block of the substance in question, 1 ft² cross sectional area by 50 ft in length, 100% insulated on the sides. The high temperature end is at 72 °F (typical winter indoor air temperature) and the low temperature end is at 30 °F (typical winter outdoor air temperature). The time stated is that necessary for 1 btu to travel from the high temperature end to the low temperature end.

{B} All compound characteristics are as presented in various tables from the 1997 ASHRAE Fundamentals guide. Thermal Dendities and Conductivities for the Earth's crusts are proportional aggregates of the physical values as documented in the Guide.

# Table B-6: Greenhouse Gas Forcing

| ppmv CO₂ | ppmv CH₄ | ppmv N₂O | % Sat H₂O | Radiant Forcing CO₂ | CH₄ | N₂O | H₂O | Total | | Line |
|---|---|---|---|---|---|---|---|---|---|---|
| 415.00 | 1.87 | 0.333 | 100% | 0.02936 | 0.00005 | 0.00002 | 0.05244 | 0.0819 | 1.9648998 | 4.1 |
| | | | | 0.09263 | 0.00015 | 0.00007 | 0.16542 | 0.2583 | 0.0061987 | 4.2 |
| 415.00 | 1.87 | 0.333 | 40% | 0.02936 | 0.00005 | 0.00002 | 0.02097 | 0.0504 | 1.2098293 | 4.3 |
| | | | | 0.09263 | 0.00015 | 0.00007 | 0.06617 | 0.15903 | 0.0038167 | 4.4 |
| 600.00 | 3.74 | 0.333 | 100% | 0.04244 | 0.00010 | 0.00002 | 0.05244 | 0.0950 | 2.2799244 | 4.5 |
| | | | | 0.13389 | 0.00030 | 0.00007 | 0.16542 | 0.29969 | 0.0071925 | 4.6 |
| 600.00 | 3.74 | 0.333 | 40% | 0.04244 | 0.00010 | 0.00002 | 0.02097 | 0.0635 | 1.5248538 | 4.7 |
| | | | | 0.13389 | 0.00030 | 0.00007 | 0.06617 | 0.20044 | 0.0048105 | 4.8 |
| 310.00 | 1.42 | 0.333 | 40% | 0.02194 | 0.00004 | 2.36E-05 | 2.10E-02 | 0.0430 | 1.0313311 | 4.9 |
| | | | | 0.0692 | 0.0001 | 0.0001 | 0.0662 | 0.13560 | 0.0032544 | 4.10 |

| btu/hour/ft² | btu/ft²/day |
|---|---|
| W/meter² | kWh/meter²/day |

**Figure B-7: Incremental U.S. Surface Temperature Increases - 1950 to 2020**

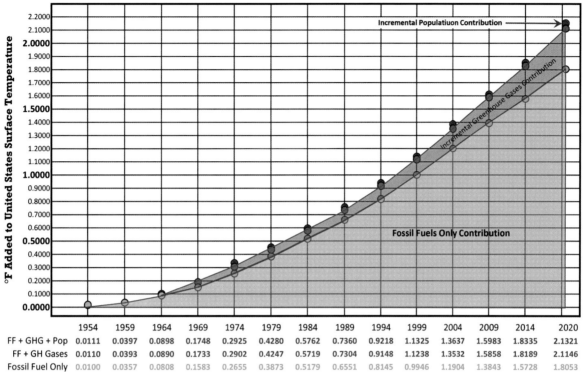

| | 1954 | 1959 | 1964 | 1969 | 1974 | 1979 | 1984 | 1989 | 1994 | 1999 | 2004 | 2009 | 2014 | 2020 |
|---|---|---|---|---|---|---|---|---|---|---|---|---|---|---|
| FF + GHG + Pop | 0.0111 | 0.0397 | 0.0898 | 0.1748 | 0.2925 | 0.4280 | 0.5762 | 0.7360 | 0.9218 | 1.1325 | 1.3637 | 1.5983 | 1.8335 | 2.1321 |
| FF + GH Gases | 0.0110 | 0.0393 | 0.0890 | 0.1733 | 0.2902 | 0.4247 | 0.5719 | 0.7304 | 0.9148 | 1.1238 | 1.3532 | 1.5858 | 1.8189 | 2.1146 |
| Fossil Fuel Only | 0.0100 | 0.0357 | 0.0808 | 0.1583 | 0.2655 | 0.3873 | 0.5179 | 0.6551 | 0.8145 | 0.9946 | 1.1904 | 1.3843 | 1.5728 | 1.8053 |

**Figure B-8: Re-Radiation from U.S. Surface with Temperature Increases - 1950 to 2020**

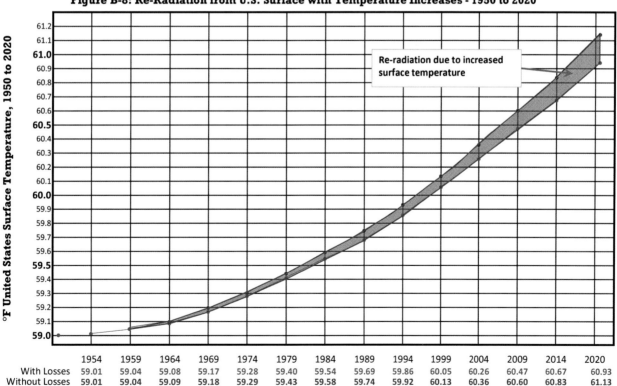

| | 1954 | 1959 | 1964 | 1969 | 1974 | 1979 | 1984 | 1989 | 1994 | 1999 | 2004 | 2009 | 2014 | 2020 |
|---|---|---|---|---|---|---|---|---|---|---|---|---|---|---|
| With Losses | 59.01 | 59.04 | 59.08 | 59.17 | 59.28 | 59.40 | 59.54 | 59.69 | 59.86 | 60.05 | 60.26 | 60.47 | 60.67 | 60.93 |
| Without Losses | 59.01 | 59.04 | 59.09 | 59.18 | 59.29 | 59.43 | 59.58 | 59.74 | 59.92 | 60.13 | 60.36 | 60.60 | 60.83 | 61.13 |

## Figure B-9: Atmospheric CO₂ Concentrations

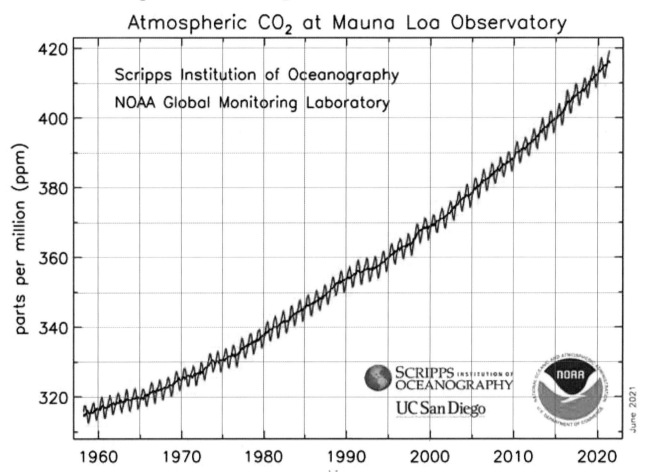

This graph depicts the upward trajectory of carbon dioxide in the atmosphere as measured at the Mauna Loa Atmospheric Baseline Observatory by NOAA and the Scripps Institution of Oceanography. The annual fluctuation is known as the Keeling Curve. Credit: NOAA Global Monitoring Laboratory

https://research.noaa.gov/article/ArtMID/587/ArticleID/2764/

*Note: the grid lines were added by the author for clarity purposes only*

# The Shiny Object in the Room: Part C

# Introduction

I have shown in Part B that the growth of $CO_2$ and the other GHGs in the atmosphere is not the reason the Earth is increasing in temperature. It's the heat, don't-cha-no! The causes for the temperature increase are both long-term and short-term. Our goal here is to create fertile ground for a rational, adult-oriented discussion, on how we should proceed into the less-than-100% fossil fueled world. Fundamentally, this is to be viewed as a point of departure, not the end-all to be-all.

The short-term cause is both the energy that we expend to run our daily lives, regardless of the source, and the sum of human bodies as a whole. The growth of humanity does two things to the heat balance. Humans release about 365 ton-hours (12,000 btu/day, 4.38 million btu/year) of heat per year, per person, into the atmosphere, and thus onto/into the surface of the Earth. Second, each human in order to maintain their existence, marginally consumes a great deal of energy just to maintain a lifestyle and a comfortable living/working environment. This is apart from the infrastructure needs like manufactured goods, roads, bridges, etc.

The long-term cause of temperature increases within our environment which I have touched on, and the explanation thereof, is beyond the scope of this document, but must be recognized and accepted. It cannot be ignored, as it has been with the IPCC. There are forces that are totally beyond our control, enumerated in **Part B**, that are moving us toward a greenhouse world. Currently, we are in an icehouse age, but given the fact that we don't have glaciers creeping up to our front door doesn't mean it isn't true. The sooner we take that fact on board and deal with the real problem, the rational stewardship of our fossil fuel resources, the better we can sleep at night.

Okay, we're going to approach this issue from three different standpoints. In **Section 1**, we will have a high-level discussion pertaining to the where we currently use our sources of fossil fuel. To some degree, this is a review of much of our discussion in **Parts A & B**, but much more focused on the probability of moving usage from the "fossil" arena to the employment of more renewable sources.

In **Section 2**, we are going to go much further in depth as to what uses of our fossil fuel resources; 1) we can't provide a substitute for, b) we can do without, and c) we can realign via the use of electricity. All of these points are critical because the solution to each of these questions will no doubt change our way of life. The best way to accomplish that change is through the use of our free-market economy and allowing the free enterprise system to do its job.

**Section 3** will take a look at both the practical and economic side of the alternate ways that we can employ to provide the electricity needed for the substitutions outlined in **Section 2**. There is a balancing act here that is not immediately obvious. Not only are we removing heat from the Earth's surface by eliminating portions of our fossil fuel diet, but if we use the Sun's energy directly for the conversion to electricity, we are removing that quantum of solar power from the functions it currently is performing. An energy balance: Energy In = Energy Out. We will talk at length on this issue.

**Section 3** will take advantage of the model that I built for **Part B** and use it to demonstrate the future impact on the globe of the projections we make for how, when and by what means we abandon the use of fossil fuels while still trying to maintain a reasonable life-style. I will project a reasonable timeline for each area of usage that will, in my estimation, impose the least damage to our economy, and the least disruption to our life style. Again, I am only going to focus on how we, within the confines of the United States, are affected. As I have stated in **Part B**, our cost of doing business has no place to go but up. Our job is to make sure it doesn't get out of control.

# Section 1: Where Does All That Fossil Fuel Go?

**Tables A-5, A-6 & A-7** from **Part A** show the proportionality of our sources of energy and how they are converted and transported to us for our end use.

## Table A-5

The Non-Renewable sources basically come from planet Earth directly and obviously need to be mined, extracted or refined in order to be converted into a commodity that can be transported via the various energy carrier infrastructures. The energy associated with this extraction/refining process is to a great extent "buried" in the usage categories for each of the forms. For instance, when we get to where electricity, natural gas, coal or petroleum is used commercially, much of the combined heat and power (CHP) segment goes to the conversion process.

Renewable sources are pretty much stand-alone and the "losses" if we can call them that, simply manifest themselves as efficiency-of-conversion during the transformation into the energy carrier infrastructure. Most of the Renewables are heat directly suppled to the end use as heat, or most likely as electricity entering the energy carrier network.

The transformation from the source to energy carrier status is quite energy intensive, especially when we address the conversion to electricity. The refining of petroleum is energy intensive, but most of the losses are part of the usage when viewed as CHP utilization.

Once the raw energy has been converted into a form that can be handled and distributed by the energy carrier infrastructure, the transportation losses are well defined. At the end use, all of the energy supplied is assumed to accrue to heat provided to the Earth's surface or given up to the atmosphere which will then be transmitted into/onto the Earth's surface. Remember, that heat has nowhere else to go but into/onto the Earth's surface as "stored" energy.

## Tables A-6 & A-7

Nuclear is the only energy source that converts 100% from the raw form to electricity. This simply means that of the energy derived from the fission process, 100% is devoted to the generation of electricity. However, the conversion process is not 100% efficient. Only about 35% ends up as electric power while the remaining 65% accrues as losses to the Earth. Currently, nuclear fuel has no other means of conversion beyond the production of electric power. Therefore, as we go forward, you will see that we are going to have to rely more and more on nuclear fuel to provide the necessary function of base-load electricity production currently performed by coal plants. Please keep in mind that this is a process that adds heat to the Earth, but no $CO_2$. Therefore, the temperature of the Earth will still increase.

The conversion of fossil fuels to electricity only consumes 27.9% of the total usage. Petroleum can be removed from electricity generation relatively easily, to be replaced currently by one of the other fuels. The obvious area to attack in order the get the biggest bang-for-the-buck through the reduction of petroleum usage is transportation. Other uses as outlined in **Section 2** are very difficult to replace with electricity; such items as lubricants, feedstock for plastics, space heating and steam generation, etc. come to mind. The biggest categories that will be almost impossible to touch are aviation, aquatic and rail transportation.

These are biggies as you will see. There are some areas of hope, however, but they are few and far between. The object in the long-term is to make sure that whatever usage we put petroleum to cannot easily, or in current economic terms, be economically replaced. As with the other fossil products, we need to preserve petroleum as much as possible for future generations.

However, in the short-term, it will be very difficult to make a rapid transition from the use of either coal or natural gas to generate electric power on demand. Right now, coal is almost indispensable as the base load plant fuel. It is cheap, readily available, and the plants can be made quite efficient and almost pollution-free. The problem that we face is the misguided perception that it is the $CO_2$ generated by coal, in quantity, is the cause of the Global Warming hoopla. Once the Green New Dealers (**GND**ers) can get over the hump that it is the heat, not the $CO_2$, and that more heat is not the cause of society's woes, then we can worry about the long-term phase out of the coal plants into either renewable sources, or nuclear. Renewables have their own problems which we will discuss, but when the transition is made to nuclear, the heat transmission to the Earth remains the same. We want to preserve the coal, not to get rid of that extra 65% that is heat loss to the Earth, but because there are uses for which we will need that fossilized carbon in the future that are <u>not</u> easily replaced.

On the other hand, natural gas is, as stated by the Wizard of Oz, a horse of a different color. It spans a huge spectrum of uses that we will go into in depth in **Section 2**. The bottom line is that in the short-term, it is almost indispensable as the fuel of choice for peaking plants serving all of our electric grids. Note what happened with the Texas Grid in the winter of 2020. They had moth-balled some peaking and base load plants and it took quite a while to get the grid operational once the windmills froze up. Right now, natural gas is cheap, marginally produces less $CO_2$ than either coal or petroleum per btu, but is very inefficient when used in internal combustion engines or turbines to power electric generators. The waste heat added directly to the Earth, as a function of kWhs produced, is relatively substantial when viewed in the light of coal or nuclear base load plants.

The critical factor when addressing natural gas usage is not electric power generation, it is its use as a primary heating fuel for commercial and residential buildings and feed stocks for manufactured products. The usage of coal for these purposes is almost non-existent except in the rare, unconverted industrial setting. Petroleum used as a source of fuel in oil-fired boilers is still employed, though probably being phased out in light of the less expensive and more readily available natural gas.

Renewable energy sources are almost totally devoted to the production of electric power, except for biomass. Only 9.4% of the biomass usage is used to produce heat which is converted into electric power. The rest is used directly as a source of heat, basically in industrial settings, such as those devoted to the lumber industry. Regardless, all that energy ends up as heat, $CO_2$ and water vapor all of which accrue to the Earth. Just to set your mind at ease, no matter how the biomass is employed, it will end up as heat accruing to the Earth. The biodegradation process releases just as much heat as combustion, and just as much $CO_2$; it just takes longer.

# Section 2: Let's Look at the Problems Facing Us When We Replace Fossil Fuels.

The first area to attack is the petroleum side of the coin, primarily because its usage is pervasive and touches all aspects of our lives and our economy. Substitutions here probably will impact our daily lives the hardest. *Table A-9* from **Part A** shows the distribution of use for all the derivatives of the crude oil that we extract from the ground and refine. 83% of the petroleum, once refined, is used for transportation purposes, even the HGL (Hydrocarbon Gas Liquid) segment (LPG, Liquified Petroleum Gas) is used to propel such vehicles as golf carts.

Take a look at *Table C-1*. This examines the addition of $CO_2$ to the atmosphere, when viewed as a contributor to the GHG "umbrella" discussion. Note that numerous sources, footnoted in **Parts A & B**, conclusively state that only about 50% of the $CO_2$ which enters our atmosphere initially, remains in the atmosphere for any appreciable length of time. The remainder is either consumed as part of the photosynthesis process (an on-going need), or is dissolved in all waters, fresh or saline, to be used by plants and animals that reside therein.

Okay, if nothing else, this should demonstrate that there is one boatload of air acting in our behalf circling the Earth. We have 5.81 billion $10^6$ tons, 5.81 quadrillion, tons of dry air above us all the time. Remember, few molecules of our atmosphere can gain enough energy to leave Earth's gravitational pull, even hydrogen. The purpose of this tabulation is to show the impact of the usage/combustion of our fossil fuel reserves on an individual basis, rather than lumping all usage into a single category.

This table throws a lot of information into the mix, so let's go line by line. Bear in mind that the conversion factors used are generic and variances occur via real-time fuel mixture decisions, types of Coal consumed, etc. Any resource that we consume whose use requires conversion into heat, regardless of the end-product, is shown here.

The first column shows the net contribution of $CO_2$ into the atmosphere per Qbtu of resultant energy. Each of the compounds are hydrocarbons, even Coal to an extent, thus not only is $CO_2$ a resultant, $H_2O$ and in many cases Sulphur and Nitrogen compounds also accrue in minute quantities. The least offensive fossil fuel of course is $CH_4$, followed closely by any petroleum product. Biomass is assumed to be the $CO_2$ equivalent of Coal. The values assigned to each fuel were derived by comparing the products of combustion to the raw material on an atomic mass basis and employing the known heats of conversion.

The U.S. and World consumption of each of the fuels is as shown previously. It shows that the U.S. consumes about 15% of the total, world-wide usage of all combustible material. However, because our 2020 consumption of Coal is so small by comparison, we only contribute 5.6% of the total heat from Coal and 13% of the total $CO_2$ into the atmosphere. Our contribution of $CO_2$ in 2020 was 0.72 ppmv, whereas the world-wide contribution was 5.62. Data shows that instruments recorded a 3 ppmv increase in $CO_2$ levels during 2020, which correlates nicely with the postulate that 50% of the $CO_2$ once it enters the atmosphere is dispersed quickly to the usage mentioned above.

Note that this analysis does not take into account $CO_2$ added due to animal respiration, though this can be quantified. That said, any $CO_2$ given up through respiration is nothing more than recycling $CO_2$ from the atmosphere, through plants and the subsequent digestion thereof. Therefore, quantifiable $CO_2$ contributions are simply delayed by some finite amount and are part of a bias that can be ignored.

We will examine in depth why worrying about the $CO_2$ contribution, as opposed to the heat added to the atmosphere which produces $CO_2$, is not and should not be where we focus our efforts going forward. Again, the real question is what will change, weather-wise, as a result of our current lifestyles, and is it going to be good, or is it going to be bad? It could result in being a "matter of taste."

Let's consolidate some of the data from *Table A-9*, **Part A**, and *Table C-1* and see what we gain and what we lose by converting some of the fossil fuel usage to electricity. I will only focus on the infrastructure constraints and/or possible changes as they relate to the U.S. Given that we are a dynamic society, and we have a bunch of smart people with an entrepreneurial spirit, I'm only going to limit our substitution discussion to the "short-term," which is about 15 to 20 years.

Also, very importantly, I will only base projections of any change on either past data back to 1950, or current 2020 data for the U.S. and only the U.S. I will not attempt to use a crystal ball and project growth of either energy usage or population, U.S. or World. So much can happen in that timeframe, to try and predict any trend is like living in a fool's paradise. However, I will go out on a limb and suggest that there are two areas of petroleum usage that can be converted, with some pain, to the use of electricity, given what we know and produce today.

## Petroleum Usage

- I think we can eliminate the following categories from any conversion prospect, using 2020 data from *Table A-9*.

|   | | |
|---|---|---|
| ○ Asphalt and Road Oil | **0.831** Qbtu | |
| ○ Aviation Gasoline | **0.020** | |
| ○ Propylene | **0.389** | |
| ○ Jet Fuel | **2.237** | |
| ○ Kerosene | **0.016** | |
| ○ Lubricants | **0.223** | |
| ○ "Other" (see table) | <u>**2.396**</u> | |
| ○ **Total** | **6.112** – 18.2% of total Qbtu/year | |

- **Area 1:** The first area we can begin the substitution process with has to do with substitutions not necessarily with electricity. It might be possible to replace these uses with either coal or natural gas, but that would only be done to preserve the reserve of the crude oil feedstock, not to eliminate heat or pretend to have an impact on GHG emissions.

|   | | |
|---|---|---|
| ○ HGL – Other | **2.387** Qbtu | |
| ○ Petroleum Coke | <u>**0.582**</u> | |
| ○ **Total** | **2.969** – 8.9% of total Qbtu/year | |

- **Area 2:** Finally, we come to the biggy; transportation and heat production.

  - ○ **Heat Production**
    - ▪ Residual Fuel Oil     **0.499** Qbtu
    - ▪ Propane     **1.142**
      - ○ Total     **1.641** – 4.9% of total Qbtu/year

  - ○ **Transportation**
    - ▪ **Distillate Fuel Oil**     1.956 Qbtu
    - ▪ **Motor Gasoline**     14.855
      - ○ Total     **22.811** – 68.0% of total Qbtu/year

**Area 1** is questionable having to do with either $CO_2$ emission reductions, or energy savings by switching to other, more plentiful fuel sources, or electricity. The free market may determine that the cost/benefit ratio for this type of conversion is just not feasible. Therefore, going forward, we will assume no change will take place in that area of concern. This leaves **Area 2**, Transportation, which consumes 68% of the petroleum that we processed in 2020.

Distillate fuel oil is used primarily in diesel engines. Over-the-road trucking, Military vessels and vehicles, the cruise/aquatic transportation industry and rail transportation consume the vast, almost 100% of this type of fuel. Each of these uses pose different problems when it comes to any conversion. The easiest, though not necessarily the least expensive, is to electrify our entire rail system in order to get rid of the diesel engines driving the generators within each locomotive. Converting large tractor-trailer rigs to electric motors is possible but might prove to be less than cost effective. These vehicles are weight-sensitive, and batteries weigh a bunch. Cruise ships and aquatic transportation/freight vessels have no chance of converting away from diesel fuel. We will discuss some of these conversions in **Section 3**.

Last, but certainly not least, is the area of personal transportation which consumes well over 95% of the automotive gasoline. The remaining usage would encompass such areas as small engine propulsion (lawn maintenance, emergency generators, motorcycles, etc.). Though possible to some degree, converting pleasure aquatic craft to electricity as a means of propulsion might prove to be politically verboten. Again, in **Section 3** we'll take a shot at some percentage reductions in each of these areas as it pertains to the substitution of electric power. The last thing we need to think about is regulating the usage of gasoline to specific consumption categories. Talk about politically verboten!

## Coal Usage

Coal is enormously abundant both in the U.S. and in the world. Reviewing from *Table C-1*, we see that the total world-wide usage of coal is 154.45 Qbtu, of which the U.S. consumption totals 9.18 Qbtu, or 5.6%. Obviously, the rest of the world, most notably India and China, have little if any concern about the use of coal as it relates to the production of $CO_2$. (Is it possible they know more than the **GND**ers?) Most importantly, I seriously doubt that they are going to do anything meaningful to reduce their usage. It is fueling a robust economy in both countries, so why change?

That said, if in fact the **GND**ers can get over the fact that $CO_2$ is not the problem and it really is the heat generated, then maybe we can begin to take advantage of our own resources and free up other fossil fuels from the task of generating electricity. We can build/un-mothball coal generating facilities a great deal faster than we can construct nuclear generating plants. We will go into those trade-offs in much greater detail during our discussion in **Section 3** because as you will see, we are going to need a bunch of electric power. Aside from the use of coal as a source of heat to generate electricity (89.65% of total usage), you can see in *Table A-12*, **Part A**, that what remains is only 10.35% of our coal production, which is used for commercial or industrial purposes. The production of Coke, 4.51%, really can't be replaced with electricity. The CHP category might have a chance to be converted to electricity (2.29%), but not the non-CHP usage (3.56%) because is it either manufacturing feedstock, or converted into hydrocarbon gas and liquid form to be used for other industrial purposes.

## Biomass Usage

The use of Biomass for whatever purpose within the U.S. and in the world is highly fragmented and very difficult to pigeon-hole into quantifiable categories. Reviewing from *Tables C-1, A-6 & A-7*, we see that biomass contributes 4.53 Qbtu, or 4.9% of our total energy usage. Electricity generation uses 0.424 Qbtu, which is 1.2% of energy used to produce electricity, or 0.5% of retail energy sales. The remaining 4.11 Qbtu is used in such areas as heat for steam production or kiln use in the lumber industry, residual incineration at landfills, or in the food processing industry. Fundamentally, it's a "throw away" hydrocarbon that is used where it can be, but otherwise would biodegrade into heat, $CH_4$ and $CO_2$. My guess is that trying to make a case for the substitution of electricity to provide the same function would be a waste of time. Again, if we can get over the $CO_2$ avoidance issue, biomass is just fine where it is.

## Natural Gas Usage

Apparently, natural gas is "everywhere" and can be accessed very easily. Currently, we have a Federal Government Administration that is doing all in their power to thwart our ability to explore for new deposits of natural gas. However, the odds are this demeaner won't last forever, simply because as soon as the politicians can't get air conditioning or heat on demand, they will probably wake up and in the customary, knee-jerk fashion, do something to correct that problem. Given our built-in system of checks and balances, within several years, we should be back to normal and worried about something other than yesterday's baseball scores. 'nuff said.

Referring to *Table A-11*, **Part A**, you can see the various categories of usage for natural gas. As you probably noted, of all the fossil fuels, natural gas, $CH_4$, is the least offensive when viewed in the light of it being a contributor to the GHG community. The reason is in the formula, $CH_4$. All other liquid or gaseous hydrocarbons have higher carbon, C, to hydrogen, H, ratio. The conversion formula is as follows:

$$1 \text{ mol } CH_4 + 2 \text{ mols } O_2 \longrightarrow 1 \text{ mol } CO_2 + 2 \text{ mol } H_2O + 891 \text{ kJ/mol}$$

This is a pretty simple equation and process. Nothing nasty results and the only products are heat, $CO_2$ and water, $H_2O$. A mol is a group of molecules containing $6.02214076 * 10^{23}$ (the Avogadro Number) molecules of that compound. The weight of both the $CH_4$ and the $O_2$ in this equation is 16.043 + 31.998 * 2 grams respectively for a total of 80.039 grams. The resultant product is 44.009 grams of $CO_2$ and 2 * 18.015 grams of $H_2O$, for a total of 80.039 grams. 891 kJ/mol of $CH_4$ gives us a heat of combustion for $CH_4$ of 23,877.16

btu/lb., coupled with 2.743 lb. of $CO_2$ per lb. of $CH_4$ consumed. Basically, we get a lot of "bang for the buck" when we use $CH_4$ as a source of heat.

On with the show. Let's break down the uses of $CH_4$ as we did with petroleum into the usage areas that we employed before. The first group would be those uses that we should not consider as substitution fodder. In **Area 1** we can do some substitution but not necessarily with electricity, and in **Area 2**, the substitution can be performed with electricity.

- Using the data from *Table A-11*, I think we can eliminate the following categories from inclusion in the substitution "locker." We have to assume that regardless of how much $CH_4$ we use, it will take about the same amount of energy to find it, bring it out of the ground, and transport it to the retail locations. That said, I think it will be very difficult to eliminate, or even throttle back, the usage of $CH_4$ in the peaking plants serving our electric power grids, at least during my short-term investigation period of 15 to 20 years. Industrial Non-CHP is used for hydrocarbon feedstock and it is possible to replace it with petroleum products. However, I'm going to consider it essential as taken out of the ground and therefore, no substitutions can be made.

  - ○ **Lease and Plant Fuel**      1.897 Qbtu
  - ○ **Pipeline and Distribution**   0.960
  - ○ **Industrial Non-CHP**        7.145
  - ○ **Electric Power Generation**  12.046
    - ○ Total              22.048 – 69.75% of total Qbtu/year

- **Area 1** usage categories are again fragmented and possibly buried within some broader classifications. Probably there is no question but what the Industrial CHP usage can be replaced, in the short term, by using either distillate fuel oil or coal. My guess is that the conversion process will be difficult, but not impossible. To try substituting electricity, on the other hand, could prove somewhat of a challenge. Quite possibly the ripple effect would work its way all the way to the sub-station level and be very expensive.

  - ○ Industrial CHP           1.418 Qbtu
    - ○ Total              1.418 – 4.48% of total Qbtu/year

- **Area 2** usage is relatively easy to deal with. The use of $CH_4$ as vehicular fuel is focused primarily on commercial rather than personal vehicle usage, basically because of the convenience of the local gas pump. That said, the current fleet can eventually be replaced with electrical vehicles of sorts when the fleet reaches its product life limit. Many urban busses currently employ $CH_4$ as a fuel, primarily due to political pressure rather than economic need. Regardless, it would not be out of the question to replace entire urban fleets with overhead, DC electrical power and resurrect the propulsion of old.  This transition of course is a blanket proposition that covers petroleum substitution which we will discuss in **Section 3**

- The residential and commercial usage has to be considered as providing heat for space conditioning purposes, or for such uses as energy at the stove-top, deep fryer, or in ovens.  No matter where it is applied the odds are almost 100% that it can be replaced with electric power.  This is not to say that the transition will be seamless, or "plug and play."  There are some substantial infrastructure problems that we will address in **Section 3.**

&#9675; **Residential**    **4.820 Qbtu**
&#9675; **Commercial**    **3.263**
  &#9675; **Total**    **8.083 – 25.57% of total Qbtu/year**

# Section 3: Energy Balances and Substitution Problems

I have used the term "Energy In = Energy Out." This is a fundamental concept learned by all first-year engineering students, but obviously was ignored when the investigators who reported their findings to the IPCC did their studies. Only recently, has the IPCC acknowledged that $CO_2$ has a lesser impact on the increase in temperature of the surface of planet Earth than previously reported. Therefore, for the last 10 - 20 years, our "climate change" discussions would have been substantially different, and we would have employed less draconian measures in the hope of reducing the temperature rise.

Again; Temperature Increase Bad, Status Quo, Okay? I still contend above all else, that to date, I have seen no evidence of a validated study that can tell us why an increase in the surface temperature of the Earth, caused by our human activity, is harmful. What is going to happen that any of the "investigators" will bet their next paycheck on? I'd love to hear the answer to that question. The odds are, not one of the investigators are willing to lay their reputation on the line by trying to predict something that they know darn good and well can't be predicted with even a modicum of accuracy. The real issue that we should focus on is good stewardship of our natural resources. But I repeat myself.

We need to re-visit *Figures B-7 & B-8* in **Part B** and the explanation for them. Remember that I used the model that I described in **Part B** to generate the data for these two Figures. These are not true energy balance graphs. Their purpose is to show incremental temperature changes at the surface of the Earth, within the confines of the U.S., that are a direct result of the increase in energy usage and population increases from the end of 1949 through 2020. These are not projections, they employ actual data as presented by the Federal Government Census Bureau, and the Energy Information Administration. Again, these graphs show only the incremental effects of the increases in GHGs, fossil fuel usage and population. However, to get a real grasp on the magnitude of the energy transfer onto and off the surface of the Earth, a true energy balance must be performed.

*Figure C-2* shows the results of this energy balance study of only the geographic confines of the United States, including Hawaii and Alaska. The first thing that should strike your attention is scale of the energy totals. The surface of the U.S. will absorb about 17,100 Qbtu net per year from the Sun. The derivation of that value is documented in *Table B-2*, **Part B**. The U.S. net TWh/year solar insolation is assumed to be 5,011,737, which converts to 17,100 Qbtu using 293.083 TWh per Qbtu. The reason Qbtu was used as the scaling factor simply had to do with size of the numbers and readability.

The assumption here is that the solar insolation is something that we have no control over, and it will change randomly day-to-day, week-to-week, but over the period of 5 years, the time step in the model, it will average out. This value in the graph is shown in **Red**, both in the bar area and the raw data. The radiation of energy from the Earth to the sky dome and eventually to outer space is driven solely by the $4^{th}$ power of the differential between those two spectrums. In other words, when the radiant energy starts its journey outward, it thinks its headed to "infinity and beyond," that great phrase from "Toy Story." Thus, unless those quanta of energy are stopped for any reason, they will keep on going and be lost forever from the Earth. This gross radiation effect is shown in **grey**, both in the bar section and the raw data.

Well, as we know, the GHGs stop some amount of that energy and re-radiate it back to Earth to some degree. This re-radiation effect is conservatively shown in green and orange, the green portion above the **blue** dotted line and the orange portion below that line in the middle section of the graph. Note that there are obvious scale differences between each of the three sections of the graph. The bottom section of the graph with values in orange demonstrate the accumulated absorption of energy by the Crust. This accumulated energy causes the temperature of the surface to increase, but over the 70+ years of the model, this increase is less than 1 - 2 °F. As stated in **Part B**, the greatest increases take place in the densely populated urban areas, which almost become permanently higher in temperature than the surrounding countryside.

In order to achieve equilibrium on the Earth, the amount of energy leaving the Earth should equal the energy given to the Earth, either through the solar influx, or the other factors such as human activity or possible re-radiation from the lower Earth's atmosphere. I assumed that the Earth "system" was operating at steady state at the end of 1949 and the total Energy In was equal to the total Energy Out. Thus, any change in that balance in 1950 was assumed to accrue to/from the thermal mass/capacitance of the Earth's crust. When the energy became part of the crust, the surface temperature had to increase for the next time step of the model.

For every 5-year increment, there either was an increase in the surface temperature, or a decrease depending upon the average of the change in the inputs from energy usage and population growth. You will note that in the 5-year sequence around 2010 and 2020, energy was actually extracted from storage in the Earth simply because the use of energy changed, the population did not grow as much, and the higher temperature of the surface forced more energy out than it accumulated.

As I have stated before, this type of model is highly sensitive to the size of the thermal reservoir, or capacitance of the Earth's crust. When I performed the combinations and permutations, it was quite obvious to me that the thermal density used for the incremental model shown in the above *Figures B-7 & B-8* was too small for this model, so that factor was increased threefold. The overall thrust of the model was not impaired, the only difference being that the surface temperatures tended to be lower than those shown in the two figures mentioned. This in fact made the results more conservative and less dramatic. Had the time steps used been 1 year, or possibly six months, the anomalies would never have shown up due to the smoothing brought about with the higher granularity.

Other than to lend a little more credence to the following discussion, I felt that it was important to get some absolute energy values in play so that the scale of the miniscule amount of energy that we have control over comes into focus. The human activity factor ranges from 0.021% to 0.059% of the total solar insolation. It was this knowledge that led me to believe years ago that something was wrong, drastically, with all of the hue and cry pertaining to "Climate Change" and "Global Warming." Obviously, the CERES project has well confirmed that the surface temperature is climbing. The issue has always been why, not if, and to continue, couple that thought with a little "so what?"

In **Section 2** above we talked about the areas of fossil fuel usage that could be altered in some way, vis a vis, substitution using another fossil fuel possibly of greater abundance, or if the substitution could be electric power. If we elect to substitute electric power for a particular fossil fuel, the issue really is how we generate that power. If we substitute one fossil fuel for another, what we gain is the preservation of the original fuel at the expense of the substitute. Depending upon what fossil fuel is substituted for another, the marginal gain in the reduction of heat to the surface of the Earth is probably minimal at best and can be ignored for our purposes. However, substituting electricity is not so straight forward.

First, the heat energy part of the balance equation will be affected more if the electricity is generated using renewable sources as opposed to using fossil or nuclear fired power plants. Second, the increased use of electricity would force some bodacious impacts on the local and regional transmission and distribution infrastructure currently in place regardless of how the additional electricity is generated. Third, the big player in the substitution equation is transportation, specifically personal automobiles. Finally, what to do when the well runs dry?

## First is the use of renewable energy sources.

This is a very complex equation to master, primarily because the biggest source is energy directly from the Sun. The collection tools to capture direct solar power are very expensive when viewed both as an initial investment and as a long-term maintenance and replacement expense. At best, the reliability of the intensity of Earth-level solar insolence is cupreous in nature and any reliance upon it as a source must be tempered with some form of flywheel. Currently, that flywheel is envisioned as perhaps batteries and/or possibly hydroelectric pumped storage similar to the example that I mentioned in the section about Dominion Power and the Bath County project. Quite honestly, the flywheel problem may pale by comparison to the need for vast amounts of horizontal real estate, situated in optimum geographic areas, and sited in such a way as to not interfere with society's social and environmental requirements.

- Another sidebar is required. Go back to *Tables A-16 & A-17* in **Part A**. If we assume that we could commandeer land at a cost of say $3,000 to $10,000 per acre[59] for Florida farm land, to $15,000 to $30,000 per acre for Florida unimproved residential land, then this translates into $1,920,000 (best case) to $19,200,000 (worse case) per square mile. In Table 16, you see that in the ideal case, we would consume 118,939 to 35,682 square miles to site enough solar panels to replace the energy stated, depending upon the density factor. Doing the math, this is a bandwidth of $2.3 Trillion to $68.5 billion to simply purchase, let alone develop the land.

Fragmenting the distribution of both the source of the power and the flywheels should be thought through thoroughly. We have a lot of roof area in our metropolitan communities, but not all of it is ideally suited to optimize the conversion of the Sun's energy into electricity. Also, to be reliable, the flywheels need to be situated at each location. Given that this methodology will reduce substantially any economies of scale from both the first cost and maintenance standpoints (repair of roof leaks?), this approach needs some very close scrutiny before it is advocated for a broad platform.

In **Part A**, *Tables A-16 & A-17*, I postulated that in order to replace the automotive 22.4 Qbtu of transportation usage, we would need to site between 20,000 and 60,000 miles² of solar panels. This is about the area of West Virginia, up to the area of Iowa. To duplicate the need using wind technology would require about 40,000 miles² which would take up most if not all of Indiana. However, the good news is that if we used either solar or wind as the renewable source, we would not add the waste energy from fossil or nuclear generation, coupled with the electrical energy itself, onto the Earth's surface. Essentially, if either of these two sources are used, the Sun's energy is a pass-through.

Now, the only real question that is left before the house pertains to the law of unintended consequences. Keeping in mind that we are dealing in bean-bag effects here, what happens when we remove from the surface of the U.S. via absorption by solar panels, the Sun's energy that otherwise would have been used for photosynthesis? How about the effect on wind and weather patterns that are driven primarily by the

Sun's energy, locally or regionally?  What about direct, surface moisture evaporation and the effect on cloud densities?

The energy that we would pull out of the wind via the use of wind generation technology would have done something. Will it disturb wind generated pollination? The agricultural industry would get mightily upset if that were the case, so how would we mitigate that effect? Would the interruption of normal wind patterns cause more cities to experience inversion layers which trap pollutants such as is currently experienced in Denver and Los Angeles?  The "not in my backyard" syndrome will be pervasive.  Currently, we have a big enough problem siting the electric power transmission and distribution system without throwing a whole bunch of windmills into the mix. These types of issues no doubt would have some substantial effect on the environmental impact pertaining to the siting of wind farms of any magnitude.

Windmills can pump water into pressurized vessels, or uphill, directly, as they have on farms and low-lying coastal areas throughout the world for centuries.  That said, the process is not reversable without the employment of another mechanical component.  This is doable of course and certainly in some locals it might be an appropriate application. The obvious advantage of this methodology is that it is its own system flywheel because the energy does not have to enter the grid until it is needed, just as in the case of Bath County.

Previously in this **Part C**, I have postulated the possibility, not necessarily the probability, of substituting electricity for 32.535 Qbtu of fossil fuel usage.  Totally, this is 35.0% of our <u>TOTAL</u> usage of energy, albeit not necessarily at 100% efficiency.  If we assume that the conversion rate from raw, fossil fuel to the retail delivery of electricity is 35 to 38%, taking into account transmission and distribution losses, then we would eliminate 20.17 to 21.15 Qbtu per year added to the surface of the Earth.  In addition, it would increase the need for electricity transmission and distribution infrastructure to handle the additional 11.4 to 12.4 Qbtu of power (see below for a probable additional requirement).  In terms of TWh, this is 3,624 TWh to 3,633 TWh additional load.  Our current retail level of electric power delivered is 12.5 Qbtu, or 3,663 TWH. Obviously, we would have to double the size of that infrastructure in order to accommodate the switch from the fossil fuel world to the electric world.  To say the least, this is not a trivial undertaking.  More on this in the second section.

## Second, What Happens to the Infrastructure?

To a small extent, in **Part A**, I touched on the infrastructure needs when we change from a fossil-fueled world to one controlled via electrical usage.  Here, we're going to deal a little more having to do with what has to happen primarily at the end use and how it will help mitigate some of the flywheel problems I have mentioned previously. I've suggested that to distribute/fragment the collection of renewable energy, most notably solar-to-electric conversion, and the necessary flywheels currently envisioned, is not prudent if economies of scale rule.  Couple that with siting problems in dense, urban environments, and the problem magnifies.

The area that touches all citizens of the United States is the electric power and climate control needs of our residences.  Of the 32.5 Qbtu that I stated could be moved from fossil fuel to electricity, 4.8 Qbtu was the residential usage of natural gas, $CH_4$, most probably in the form of space heating.  That amount, coupled with the 3.3 Qbtu used commercially, actually provides up to 90% of its energy directly to the space.  This alters my statements above by really adding the need for an additional 4.4 Qbtu to the grid requirements, or 1,289 TWh per year.

Many of the residences in our northern climes use the natural gas mentioned above for space heating purposes. Some, or maybe all, don't have central air conditioning or even a furnace and duct system installed. To transition to electric heating might require substantial structural and heat distribution system modifications that are not cheap. Couple that with the possible requirement of having to increase the feeder power lines, distribution transformers, and the service entrance from 125 amps to 250 amps or more, and we're talking some big bucks. All this cost may not come directly out of the homeowner's pocket, but it will show up on their utility bill, forever and ever.

Now, in addition to the heating requirements, each residence, at least in our more temperate regions, at one time or another requires space cooling. Obviously, this is geographically dependent, but the total energy requirements throughout the year, be it "hot" energy or "cold" energy, sorta levels out. This leads to the conclusions that I was able to draw as a result of the project that I mentioned that was done under the auspices of Dominion Power during the 1980's and early 1990's. The project involved the installation of pilot versions of a whole-house energy management system in 12 all-electric residences. The systems were installed in 1983 and operated on a special demand-based rate structure; 11 in Northern Virginia, and one in Miami, Florida. Power was purchased "off-peak" at a substantially reduced rate, but during the "on-peak" periods, not only was the power more expensive, but the rate structure had a demand component that penalized the homeowner if too much power was consumed during a time-oriented, 30-minute window. The anniversaries of the window took place on the hour and the half hour.

The purpose of the study, from Dominion Power's standpoint, was to see what could happen to the residential sector diurnal profile of their generation, transmission and distribution system if an economic incentive was provided to the homeowner which hopefully caused them to move their usage away from the utility peak load periods. As I have stated previously, residential usage of electrical power is highly diversified in short time-frame windows. However, diurnally, their segment still sees substantial differences in requirements between the daytime and night-time periods. Over half of the annual usage of electric power for these residences was used to condition the space and/or heat their potable water supply, regardless of the season.

Each of the residences in question had systems that we designed, installed and maintained over a 10-year period. Each was controlled by an IBM-clone computer operating proprietary software and hardware of our design. U.S. Patent no. 4,645,908 illustrates the system hardware employed, and the algorithm used to operate the system. We gathered data on the environmental conditions, the times of usage of electric power, and the resulting impact on the power grid for a period of 10 years.

The end result of the study was quite enlightening. What I had postulated at the beginning of the study, both intuitively and through some extensive mathematical modeling of the residential environment, was that we had available built-in thermal storage of the building itself that could be utilized. Essentially, if we added a reasonable amount of controllable energy storage, and if we could know and predict power usage patterns, we could level the individual, residential loads. This individual control, when seen on a large, diversified basis, substantially reduced the infrastructure needs of the utility while still maintaining a reasonable comfort level within the residence.

The take-away from this study is that there are two distinct end uses of electric power, both commercially and residentially, but especially at the residential level. The first is the discretionary usage for such tasks as cooking, lighting, etc., which if attempted to be controlled would impact the occupant's lifestyle. The second is controllable within reason and includes the power necessary to condition the space and to heat domestic water without impacting a lifestyle.

I bring this study to the forefront in order to point out that there are more ways than one to "skin the cat." The inherent thermal masses of our building structures can be used as flywheels with little or no imposed cost penalty beyond a control system. A greater impact can be realized by adding additional thermal storage, but it does come with a cost penalty. The cost/benefit analysis must take into account rate structures and the increased cost of the infrastructure modifications required. Essentially, this is storage of energy in the "usage" form rather than as raw power that could eventually be used, such as in batteries or the pumped storage mentioned. However, there are efficiency and hardware cost penalties that need to be considered regardless of what the end use turns out to be.

No matter how the problem is viewed, if we begin to rely upon renewable sources of electrical energy, a means of riding through the "lean" times must be designed into the system. Assuming that any negatives that I mentioned above pertaining to the rechanneling of the Sun's power to either solar panels or windmills turn out not to be a problem, then we need to make sure another Texas-style ice storm does not cause some grid-wide problems that become political nightmares. The real question before the house pertains to how big the flywheel needs to be and how is it controlled. These are cost/benefit tradeoffs that need to be addressed on an individual basis.

Regardless of how the end-use issues are approached, addressed and solved, we will still have to upgrade the existing electric power transmission and distribution system. Dominion Power, and probably other utilities, already have faced the "not in my backyard" hue and cry when upgrading transmission lines. The burying of transmission lines comes at both a first cost and a long-term maintenance penalty. The size of the substations is directly proportional to the down-stream power needs. Also, no matter where the renewable power is generated, the odds are extremely high that it won't be close to the existing grid system. Therefore, a new subsystem must be built which again impacts the environment.

The bottom line is that the move to renewable sources of energy on a national scale has to be approached slowly and with a great deal of rational planning. I don't know if politics, or emotions for that matter, can be nullified, but I can hope, can't I?

## Third, What About the Automobile of the Future?

Currently, the miniscule number of EVs on the road don't even provide a blip on the electricity distribution radar screen. However, if California continues down the path it has taken, it will begin to show some severe growing pains that that Woke-controlled political structure hasn't thought of yet. The marketing arm of the EV industry, the drive-by mainstream media, and the **GND**-Democratic Party has used a statement about battery charging that will come back to haunt them. There is no way, at least within the next few years, that all the battery charging for the "family car" is going to take place at night, regardless of where that vehicle is located.

The Sun doesn't shine at night, the wind sort of dies down normally, so what we are left with as it pertains to a source of cheap power is the fossil fuel sector, something I thought we were trying to wean ourselves of. Well, we could use all that power that was stored in batteries somewhere somehow, pipe it through grid #1 to intersect with grid #2 (the current grid), get it to the residential service entrance and finally into the EV battery system. It smells to me like there are some decent efficiency losses here that need to be addressed. When reality sets in, the Utility Company that provides the power to charge those batteries, when charged at night, will be using fossil fuels to generate the electricity. Bottom line? More heat generated, more GHGs expelled, and the net energy usage increases.

We ignore human nature at our own peril. When we view the number of drivers who run out of gas on the highway and the lines at the gas pump during the day, I would say that we are of a culture that waits for the "near empty" light to shine before we do anything. Probably, in the broad scheme of things, by far, most of the recharging is going to take place during the daytime. I don't think it will take too many instances of an EV running out of battery power before recharging takes place at times of convenience.

To get an EV up and operational once the battery is totally discharged is not like dumping a couple of gallons of gas in the tank and sending it on its way to the next charging station. It's pretty tough to run out of battery, take up your empty can of "electricity" and hike to the local charging station to fill it up. I don't know how much that full can of "electricity" would weigh, but I would not look forward to using that as a form of exercise to replace a trip to the gym. Imagine the embarrassment of running out of battery and having a gas-powered pickup truck pull up, towing a trailer with a diesel-powered generator mounted on it, for the sole purpose of recharging EV batteries for drivers in distress, taking hours to do so. Wow, what a picture that would make.

EVs and their refueling depots are to the renewable energy electricity generation industry as the chicken is to the egg. Let's think: "Gas Station" syndrome. Currently, we take about 10 to 12 minutes to refuel our vehicles, assuming we don't use the pause in our daily lives to get a cup of coffee or another beverage/snack. Fully charging an EV will take 4 hours, best case, but more probably 8 hours. The batteries can be "topped off," but it takes longer than 10 minutes to do so. A car is not a cell phone that you can keep "plugged in." Many of us don't have the luxury of large parking areas to plug our EV into while we wile away time at "work." Therefore, some thought must go into how to replace the function and real estate necessary to deal with the Gas Station syndrome as it applies to EV charging stations.

Even the "work" issue needs addressing because every parking lot, parking area and garage is going to have to be retrofitted with charging stations and the corresponding tie to the internet for billing purposes. If the internet goes down, or you can't reach a cell tower inside of the parking structure, how will the recharging be handled without some very expensive infrastructure expense? What would happen if the number of EVs out-pace the number of charging stations? It will evolve into a game of "musical charging stations." We have a new saying coming up to replace "my dog ate my homework." "My battery ran out of power, so I'll be there sometime this afternoon for that 0900 meeting." Food for thought, at a minimum.

All the areas of the country have historic weather patterns that force extensive, electrical power outages that can last days and weeks at a time. At a minimum, the power must get to the recharging station regardless of how it is produced. When Irma "breezed through" Southwest Florida a couple of years ago, our house was without power for 12 days and we had to travel over 5 miles to the nearest **GAS** station to refill the car and get gas for the generator. Even then, the lines were huge for several days. What about those apples, McDuff? A portable, gas-powered generator maybe? How about that whole house, natural gas or propane powered emergency generator that you just had installed?

Okay, these are human life-style oriented concerns that need to be addressed, but to a degree are independent of the massive infrastructure changes that must take place when viewed from the Public Utility perspective. How are the state and local level corporate commissions going to handle pricing of the product? Are we going to be dealing with massive subsidies, one way or the other? Currently, all electricity usage passes through a meter before retail distribution. Even public street lighting is metered. So, is the Public Utility of record in a particular locale going to provide the metering and charging structure, or is this going to be a free market issue? Is the cost of the power entering the battery on the vehicle going to be of fungible cost, or is there going to be a premium charged? Who sets the rate? If the rate per kWh is so far and above what the standard, residential customer pays, what's to prevent the residential customer from going into the battery charging business in his/her local neighborhood?

What if the reverse is true? What is to prevent people from charging "home" batteries at a vehicle station cheaper than they can buy power at home to fire up their stove top? Due to the times of the day that renewable energy is available, is the residential customer going to face mandatory Time-of-Use rates which may or may not apply to vehicle charging? Per the subsidy question above, which direction would a subsidy go? More to charge the vehicle, or less to charge the vehicle? Most politicians forget that the general public is not stupid and will adjust their buying/usage habits along financial lines.

All the above really points to the fact that, at a minimum, in the short run, we need to recognize that recharging is going to add to the daytime peak energy usage. The only practical way to avoid that is to provide almost draconian differentials in energy cost at the recharging stations which would trigger reasonable, financially sound decisions on where, when and when not to recharge those batteries. Again, this assumes there is a location to do the recharging that has some modicum of convenience.

We will have to accept the fact that the transmission and distribution infrastructure size and cost will probably double over the next 20 to 30 years. We will see this reflected in our utility bills, most likely as it pertains to power used to charge EVs. The state and local Corporate Commissions will be under enormous political pressure to make sure that the additional cost for this additional infrastructure is passed on to the owners of the EVs and not to those residents who elect to keep their gas-powered "clunker."

Up to a decade or two ago, electric utilities were really only interested in building power plants and selling kWh's out the back door. When I was involved in the 80's and 90's with the industry, the custodian for their office buildings had a bigger office than the whole marketing department. Only recently are they bending to the pollical pressure provided by the **GND**ers to employ renewable power sources. It won't be long before they run out of room to site solar panels or windmills without running into opposition from environmentalists and the Sierra Clubs of the world. That would be quite a sight to see: The Green New Dealers at one end of the field, and The Sierra Club at the other end. Then the whistle blows and the referee yells "let the best man win!"

So far, we don't have much of a problem, as I stated at the beginning of this section. This is because the EV industry is cherry-picking customers from the Woke Tree of Woke-Life, which will run out of fruit in a short length of time. Other than some draconian mandate from either the Federal or State governments that all new vehicles must be electric, and all current gas-powered vehicles, once traded in must be destroyed, internal combustion engine powered transportation is going to be around for a very long time to come. If the politicians decide to declare that the existing gas-powered vehicles have no value, the hue and cry will be deafening. The answer? Let the free-market rule and get the Government the Hell out of the way.

## Finally, What About Those "Wells" Drying Up?

We have two "wells" that we have to worry about: the well containing fossil fuels, and the short-term well that we will create when we move to a larger and larger share of our electric power generated with renewable energy sources/hardware.

The Fossil Fuel Well is quite easily defined and of finite quantity. The only real question before the house is how big is finite and how long will we be able to dip into that resource, given projected rates of usage? Currently, extraction and subsequent usage is based solely on economic need and the maintenance of lifestyles, including that of the Saudi Royal Family and other members of OPEC. To continue usage with reckless abandon will result in global panic when the truth surfaces, and the "haves" and "have nots" will actually go to war with each other.

It is at the juncture where fuels begin to be withheld by the haves from the have nots, that the question I posed earlier becomes prominent. Who owns the fuel? Is it owned equally by all citizens of planet Earth, or owned by happenstance simply by geographic positioning of political borders? Right now, we are operating in the "fat, dumb and happy" world. As I have pointed out above, there are some functions that fossil fuels perform that cannot be replaced easily, or even at all, with solar energy or energy derived from sub-surface sources. These sources only provide heat and cannot directly create the complex hydrocarbons that were created millions of years ago under enormous pressures and temperature.

This is not to say that substitutes cannot be made or manufactured. Hopefully, if the transition away from fossil fuels takes place slow enough, the free market and good old capitalism will create fertile ground for the growth of an industry we can't even envision at this moment in time. This replacement technology, whatever it might be, is not a "no-brainer," because if it was, we would already be there; up and running. Again, we have to get the government out of the way so that unfettered thought can be devoted to product development, rather than satisfying the perceived needs generated in Washington by bureaucrats who are really clueless as to how the real-world functions. I think our experience with the first several months of the Biden Administration proves my point.

All that said, the key to the timing of this product development is transparency of information. Accurate reporting of the pending problem, and the real data pertaining to possible dry-well conditions needs to get into the hands of the general public. Our experience with the COVID-19 pandemic lack of flow of accurate information should tell us why trusting Washington to lead the way is a fool's errand. The general public is not stupid. They just need a reliable source of accurate information in order to make critical, life sustaining decisions.

The Renewable Energy Well on the other hand is going to be an on-going problem of an intermittent nature. Obviously, the Sun is not going away anytime soon, and for our purposes, it will never go away. Thus, we have to consider it as the only reliable "well" of energy to draw upon. I have spent a great deal of prose talking about the reliability of any energy from, and any energy converted from, the Sun. During the time when the Sun is at its best, we have more than we can use. As I have pointed out, that is not the issue. We can store that instantaneous energy production in batteries, but on a nation-wide scale, without a fossil-fueled backup, this will become totally impractical. When those fossil fuels go away, our options are limited.

I illustrated my experiences with thermal storage in the Dominion Power project that we participated in during the 80's and 90's. The lessons learned at that time are still valid. Basically, the best place or venue to store energy during times of abundance, to be used during the lean times, is at the point of use and in the form that it will be used. Also, the example of the Bath County storage/generation system is very valid. Fundamentally, hydroelectric production does not have to be limited to rivers and streams. I mentioned windmills and their ability to pump water uphill, to be used later for power production.

Going forward, we can't just concentrate on the generation of power from renewable sources. We have to focus on that time when the fossil fuels are not going to be there to bail us out of a "hot spot," so to speak. Are we wasting too much time now worrying about how to generate the power instead of how to store it in whatever form to be used in the lean times? As with the windmills, why generate electricity to go into a battery, to be removed from the battery to go into the grid.? Quite possibly, the efficiency of pumping water is substantially greater than the multiple conversion processes involved with getting power into and out of the battery into the grid. Most importantly, water doesn't wear out and we don't have to pay China to manufacture it.

This is but one example of the areas that we need to concentrate on going forward. First, we keep the citizens of the world informed as to when the end of the cliff could be. Second, provide them with a plan to keep us from going off of the cliff. Third, tell them what they can do to help in the process when the world begins to make the transition. But most importantly, be transparent. The last thing anyone wants to happen is to be blind-sided with a problem that had they known about, they might have done something differently.

I rest my case.

# Tables and Figures Part C

## Table C-1
### CO$_2$ (Greenhouse Gas) Contributions, U. S. and World-wide

| Weight of the world Atmosphere, 10$^6$ Tons of dry air = 5,809,373,437 | | | | 10$^6$ Tons CO$_2$ added to the World atmosphere 2020 | | ppmv CO$_2$ Added, 2020 per lb dry air whole Earth | |
|---|---|---|---|---|---|---|---|
| Compound | 10$^6$ Tons CO$_2$ per Qbtu | Qbtu 2020 | | 2020 | | | |
| | | U.S. | World[1] | U.S. | World[1] | U.S. | World[1] |
| **Natural Gas** | | | | | | | |
| CH$_4$ | 57.44 | 31.54 | 126.50 | 1,811.8 | 7,266.6 | 0.21 | 0.83 |
| **Coal** | | | | | | | |
| C | 155.07 | 9.18 | 154.45 | 1,423.6 | 23,951.0 | 0.16 | 2.72 |
| **Petroleum** | | | | | | | |
| Octane | 76.90 | 18.79 | 102.18 | 1,445.2 | 7,857.3 | 0.16 | 0.89 |
| Diesel | 78.71 | 10.21 | 55.51 | 803.6 | 4,369.0 | 0.09 | 0.50 |
| Other | 85.00 | 4.53 | 24.63 | 385.1 | 2,093.9 | 0.04 | 0.24 |
| Sub-Total | | **33.53** | **182.32** | **2633.9** | **14320.3** | **0.30** | **1.63** |
| **Biomass** | | | | | | | |
| Assume C | 155.07 | 4.53 | 54.15 | 702.5 | 8,397.2 | 0.08 | 0.95 |
| Total | | **78.78** | **517.40** | **6,571.7** | **53,935.1** | **0.75** | **6.13** |
| % of World | | 15% | 100% | 12% | 100% | 12% | 100% |

[1]2020 World data is extrapolated from 2017 World data using U.S. ratio 2020/2017. World data incorporates all U.S. data

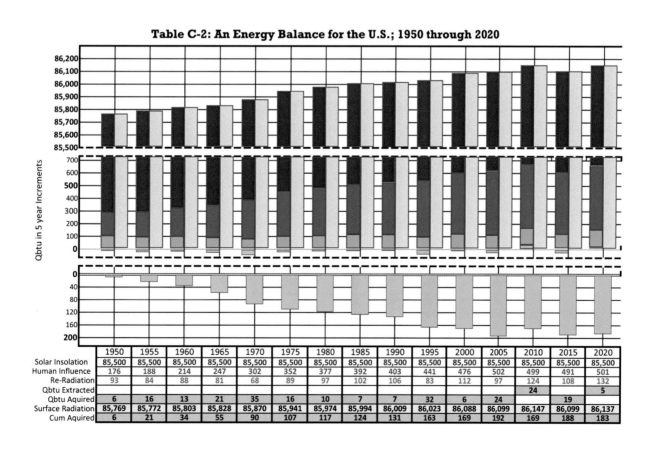

### Table C-2: An Energy Balance for the U.S.; 1950 through 2020

Qbtu in 5 year Increments

| | 1950 | 1955 | 1960 | 1965 | 1970 | 1975 | 1980 | 1985 | 1990 | 1995 | 2000 | 2005 | 2010 | 2015 | 2020 |
|---|---|---|---|---|---|---|---|---|---|---|---|---|---|---|---|
| Solar Insolation | 85,500 | 85,500 | 85,500 | 85,500 | 85,500 | 85,500 | 85,500 | 85,500 | 85,500 | 85,500 | 85,500 | 85,500 | 85,500 | 85,500 | 85,500 |
| Human Influence | 176 | 188 | 214 | 247 | 302 | 352 | 377 | 392 | 403 | 441 | 476 | 502 | 499 | 491 | 501 |
| Re-Radiation | 93 | 84 | 88 | 81 | 68 | 89 | 97 | 102 | 106 | 83 | 112 | 97 | 124 | 108 | 132 |
| Qbtu Extracted | | | | | | | | | | | | | 24 | | 5 |
| Qbtu Aquired | 6 | 16 | 13 | 21 | 35 | 16 | 10 | 7 | 7 | 32 | 6 | 24 | | 19 | |
| Surface Radiation | 85,769 | 85,772 | 85,803 | 85,828 | 85,870 | 85,941 | 85,974 | 85,994 | 86,009 | 86,023 | 86,088 | 86,099 | 86,147 | 86,099 | 86,137 |
| Cum Aquired | 6 | 21 | 34 | 55 | 90 | 107 | 117 | 124 | 131 | 163 | 169 | 192 | 169 | 188 | 183 |

# Appendix A – A Treatise on Radiant Energy

# Introduction

One of the most difficult heat transfer concepts to understand is radiation. As a matter of fact, almost every aspect of heat transfer is "mysterious" until we start to look at it from an atomic or molecular standpoint. It goes without saying that all matter, gaseous, liquid or solid is composed of atoms. Most of the time, these atoms are combined into molecules, some substantially more complex than others. Regardless of the simplicity or complexity of the molecules that compose any substance, that substance will transfer heat, either internally or externally, and to varying degrees. The mystery is how that transfer is accomplished. More than likely, that transfer takes place either through molecular-to-molecular or atom-to-atom "collision," intermolecular radiation, or possibly conduction if it can be conclusively separated from the first two. Collisions can take place with gases[60], but probably not with solids or tightly bonded and very viscous liquids. This is the process called "Brownian Motion." What we will focus on here is heat transfer through radiation; specifically, from relatively solid substances to gases, and gases to gases, though not being unmindful of molecule-to-molecule "conduction."

What we are going to show in this **Appendix** is that almost beyond question, any **GHG** that we have any hope of controlling has little if no effect whatsoever on the increase in the Earth's surface temperature. Remember, it is the Earth's surface temperature that heats the atmosphere that drives the weather. The only **GHG** that might have some effect, positive or negative, is $H_2O$ which, **a)** we have no control over, and **b)** has been around since time immemorial, as have all the other atmospheric gases. It isn't just adiabatic expansion that heats or cools our atmosphere. It's surface radiation. If that radiation was shutoff, the atmosphere would soon reach the temperature of outer space, and we would no longer have to worry about getting that broken ice machine in the kitchen refrigerator repaired.

Let's get moving because we have lots of lines, charts, graphs & arrows to look at and talk about.

# GHGs

*Table A-1* illustrates the 10 most abundant gases that compose the atmosphere of Earth[61]. Only 4 of those gases, water vapor ($H_2O$), carbon dioxide ($CO_2$), methane ($CH_4$), and nitrous oxide ($N_2O$), in order of importance, are considered GHGs (**GHG**). Essentially, they are classified as such simply because they are substantially affected, though to varying degrees, by their impact with infrared radiation[62] emanating from the surface of the Earth, and from the Sun. The surface of the Earth radiates in the infrared spectrum upward into the atmosphere day and night, therefore, the fact that the sun happens to be shining at any moment in time does not significantly impact our conversation.

Please keep in mind that all atmospheric gases are affected by infrared radiation; it is simply a matter of degree. We should mention at this point the significance of each of these **GHG** gases as it relates to the surface temperatures of the Earth that we experience on a yearly basis.

*Table AP-1*[63] [64] documents the absorption characteristics of each of the gases in question. Within the "habitable" temperatures in the table, $CO_2$ will absorb energy in any quantitative measure that emanates from a source 57 °F or less (**Red** sidebar on the right hand of the table), or minimally from ~85 °F to ~115 °F (diminished **red** sidebar). $N_2O$ and $CH_4$ likewise require radiation coming from sources in excess of 130 °F (Grey sidebar). Water vapor, $H_2O$, on the other hand absorbs energy from sources at temperatures most likely to occur in abundance on the Earth's surface (Blue sidebar). Obviously, by inspection, at <u>best</u> $CO_2$ can only accept energy, and reradiate some amount of that energy back toward Earth, less than ½ the time. By definition, only ½ of the Earth at any moment in time is below 57 °F.

*Figures AP-1 to AP-4* depict what each of these **GHG** molecules "look" like. All "other" elements that comprise our atmosphere are for the most part affected to a much lesser degree by radiant energy resulting from commonplace, Earth-side temperatures. Don't get me wrong because all "other" elements in the atmosphere contribute. As you will see, it is the contribution **per molecule or atom** that is of importance. Many of the "trace" gases are predominantly affected by energy in the ultraviolet region, and within the upper atmosphere. Toward the end, we will discuss how the "other" atmospheric gases are affected and what that effect means for us "Earthlings."

<u>All</u> gases in the atmosphere are affected one way or the other by the radiation coming from the Sun. This radiation covers the "spectrum" but its affect on surface temperature rise has already been taken into account in countless studies having to do with Solar Incidence. This net Solar Incidence, as opposed to the total of Solar Incidence, is the result of the Albedo[65] effect which is substantially more affected by clouds and surface reflection than any absorption characteristics of the **GHG** content of the atmosphere. As a mater of fact, we really don't care if a **GHG** intercepts any radiation from the sun because if it does, it has a tendency to increase the Albedo effect as opposed to decreasing it. Essentially, an increase in the albedo effect decreases the surface temperature of the Earth rather than increase it. Therefore, we really are not

concerned with what happens during the day when the Sun shines, only at night or in very cloudy conditions when there is the possibility of net-radiant loss or gain from the surface of the Earth.

## What Happens When the Sun Shines?

Time to throw a little logic and common sense at the problem. We have been talking about the theoretical effect of **GHG**s on the warming of the Earth, and by association, the atmosphere. The basic **GHG**s that are of concern, $CO_2$ & $H_2O$, absorb radiation of differing wave lengths, or frequencies (see *Table AP-1*). Essentially, the molecules in question demonstrate on a molecular level what a spring does in the "real world." If you take a weight of the proper size, attach it to the bottom of a reasonably flexible spring, hang on to the end opposite of the weight, and then "bounce" the assembly, the weight will occilate up and down until all of the imposed energy is expended. The assembly literally has a "resonant" frequency that was initiated when it started in motion. If you continue to "tap" the weight, it will keep in motion until you stop replacing the expended energy.

Okay. Visualize that a molecule of **GHG** also has a resonant frequency that is set in motion (excited) by an electromagnetic wave form, and will continue in motion if continually reinforced. OBTW, If not reinforced, it will stop until re-excited. The "excitement" generated its own wave form (re-radiation), but not necessarily at the same "excitement" level. In other words, it may be excited by a wave emanating from a body at say 4,300 °F, but its "resonant frequency" only allows it to vibrate and emit wave forms at say, 200 °F. Taking this analogy a step further, if our molecule in question doesn't become excited via some external stimulation, it remains idle, or dormant, and has no "temperature."

Let's look now at *Table AP-4*. This table tells us a great deal about what happens in our atmosphere during the daytime when solar insolence is present. Night and day, **GHG**s are subject to bombardment from waveforms of all natures; but at night, the only waveforms are those in the infrared regions. Looking at columns 11 & 12, you can see the substantial difference in the stimulation affect between night and day. In the daytime, the waveform power and the number of "hits" from the energy generated by the Sun far outweigh any contribution from those waves emminating from the Earth's surface, when the average temperatures hover around 59 °F, or 15 °C, plus or minus.

The bottom line here is that the only time we really need to be concerned about any re-radiation effects from **GHG**s is during the night-time hours and even then, as I have proven in the models, those effects are almost in the noise. It all has to do with resonant frequencies and what waveform will help, and what waveform will not help, and/or maybe even cancel the effects of other waveforms. Remember, what happens in the atmosphere during the daytime already has been taken into account via all the miriade of scientific data on the subject.

The infrared radiation that we are concerned with is classified in the region of long-wavelength infrared derived from substances within the temperature range of 192 °F to -112 °F (89 °C to -80 °C, wavelengths of ~8.0 μm to ~14.0 μm). The temperature range for the surface of the Earth we are really concerned with, from a thermal modelling standpoint, is around 59 °F, ± 10 °F, wherein the wavelength is ~10± μm (1 μm

= 10⁻⁶ meters, = 10⁴ Å, Angstroms). Going forward, you will see why dimensioning in terms of Angstroms (Å) is important.

- It should be noted here that all scientific publications of any worth, accept the 59 °F value almost exclusively, along with the equations used to quantify both atmospheric temperatures and pressure values as a function of the distance from the Earth's surface. See *Figure AP-10* for quantification.

The **photon** energy of this wavelength would therefore be 121 to 126 meV, or million electron volts[66].

1 eV = 1.6022 * 10⁻¹⁹ J (joules) = 1.5186 * 10⁻²² btu

1 TeV = (trillion eV) 1.6022 * 10⁻⁷ J, or put in practical terms, is "about the kinetic energy of a flying mosquito"[67]

The equation quantifying the amount of energy transferred between two bodies of unequal temperature is known as the Stefan-Boltzmann (S-B) Law [68] [69], which has been verified as "real" science since before the turn of the 20th century. Within the area of our discussion, this law applies regardless of the wave-length of the radiation because the derivation of the S-B Constant takes wavelength into account. Keep in mind that the absorptivity and emissivity of a body for radiation is the same quantity and is expressed in terms of percentages; 0% to 100%. A body with an emissivity ($\varepsilon$) of 90% for specific wavelengths, will also absorb 90% of the radiation to which it is exposed; the _principle_ of absorption and emittance being identical. The following is the S-B equation for the power of the emitting radiation, with dimensions of btu/hour/ft²/°R, or W/m²/°K⁴.

$$P_{emit} = A * \sigma * \varepsilon * T^4$$
Where:
**A** is the area of the radiating surface
$\varepsilon$ is the emissivity/absorptivity of the emitting surface
**T** is the temperature of the emitting surface
$\sigma$ is the Stefan-Boltzmann Constant
         = 5.670374 * 10⁻⁸ W/m²/°K⁴ - SI Dimensions
         = 1.71344 * 10⁻⁹ btu/hour/ft²/°R⁴ - IP Dimensions

Essentially, this equation shows us that a body, with area "**A**" will continue to lose energy at the rate shown unless it somehow reacquires energy from another source. The key to understanding the significance of this equation, with respect to our discussion, is to focus in on "**A**". The validity of the equation is derived from the copious amount of experimental data which has proven the science. Basically, **A** can be any flat surface, or any number of "flat" surfaces that can be differentiated such that the radiation from one of the surfaces does not affect the radiant ability of any other surface, within the group of surfaces, to perform identically as the emitting surface performs. Also, of great importance here is the assumption implicit in the equation that the source "flat surface" is an infinite "source" of energy and does not change temperature as a function of time.

To put a fine point on this portion of the definition we need to talk about really small surfaces, because this is going to be important downstream. The smallest granularity to be considered is at the atomic or molecular level of the substance in question, because that's where the "heat" is. As you will see, there is nothing but "nothing" between molecules that could be classified as, or contain, "Heat."

The "flat" nature of the surface forces the radiation to be hemispherical in nature. However, when viewed as a group of flat surfaces, each hemisphere negates the effect of the neighboring surfaces such that the emitting radiation can be viewed as a plane, propagating outward parallel to the emitting surface. Thus, until the "wave" or quanta of "photons" travelling at the speed of light, encounters "something," the strength of the emittance, or energy density, is not diminished or altered in any way.

However, if our surface in question was spherical in nature, as a molecule or atom would seem to be, it would radiate as a sphere displaying an energy density reduction proportional to the third power of the distance from the emitting body. This, by the way, is exactly what happens with the Sun's radiation. However, with a flat surface which is composed of a seemingly infinite number of spherical objects, the net radiant loss is only that which leaves the flat, exposed surfaces, and acts as a "plane-like-front" with no diminishment in energy density with respect to distance traveled. All of the radiation that emanates other than outward from the flat surface is simply absorbed by surrounding spherical surfaces, reradiated from surface-to-surface until finally an amount, quantified using the S-B equation, escapes outward/upward.

Obviously, this example only works with tightly bonded substances that retain a consistent juxtaposition, such as the solid surface of the Earth. Also, the underlying assumption is that the source of heat causing the radiation in the first place is somewhat "infinite" as a function of time. Shortly, we will illustrate what happens when the molecules in question are separated by several magnitudes of their own diameter and we bring "**time**" into the discussion.

Essentially, radiant heat is generated, and absorbed via the "vibration" or "agitation" of a molecule or atomic bond/structure of the emitting/absorbing substance (see "**What Happens When the Sun Shines?**" above). Again, view *Figures AP-1 to AP-4*, and compare them with *Figure AP-5*, the Oxygen, $O_2$, Molecule. The Oxygen Molecule is very tightly bonded and can't "vibrate" to the same degree as say $CH_4$, especially enough to absorb or emit infrared radiation of any magnitude within the wavelength and temperature ranges that concern us here. However, when in the gaseous state, $O_2$ can be "agitated" as in Brownian Motion, and transfer its kinetic energy via "collisions" with its counterparts, thus transferring "heat."

Radiant heat is transferred electromagnetically through the void between molecules and atoms until it reaches another absorbing surface, or not. In our case, when a quantum of the radiation leaves the surface of the Earth, it propagates at the speed of light (186,282 miles per second, 299,792,458 meters per second) until it encounters a surface that absorbs its quantum of energy. If it does not get waylaid by an atmospheric gas, or any other absorbing substance for that matter, it escapes into outer space and is "gone forever." As a point of reference, light takes about 8 minutes and 17 seconds[70] to travel from the Sun to the Earth. Again, for some prospective, the quantum of infrared radiation that we are focused on takes about 0.0333 seconds to reach the top of the Exosphere[71] 6,200 miles above the surface of the Earth, assuming it doesn't "run into anything."

Okay, we need to spend a little time on that term "void," or possibly "nothing." Our atmosphere, regardless of elevation above the Earth, is almost totally composed of "void." Take a look at *Table AP-2* which enumerates several key parameters pertaining to the gases that compose our atmosphere. As mentioned, there are 10 primary gases that are contained in our atmosphere, coupled with some gases that are "trace" in nature, such as some left-over halogens, carbon monoxide (CO), and ozone ($O_3$), that exist mostly above the tropopause and have little to do with a greenhouse effect. **Lines 1** through **10** provide information pertaining to the important characteristics of these 10 gases as related to their effect on our lives.

- **grams per ft³** are calculated knowing the ppm by mass which is derived from the ppmv shown in *Table A-1* and the known atomic weight of each molecule.
- **mols per ft³** are calculated knowing the molar weight of the element. This then allows us to calculate the number of molecules per ft³, using Avogadro's Number.
- The atomic and molecular diameter, using Angstroms (Å) as the dimension of choice, is derived from known chemical data. Molecular distances are normally stated in picometers (pm), as you can see illustrated in *Figure AP-4* for nitrous oxide. 1 Angstrom, **Å** = 100 pm.
- Knowing the molecular diameter, the enclosed, spherical volume can be calculated and expressed in $Å^3$.
- **Line 13** shows that the total volume of 1 cubic foot of atmosphere expressed in $Å^3$ is = $2.832 * 10^{28}$. Also, on this line is the total percentage by volume of all **GHG**s, 0.0000845%.
- The last column shows the percentage by volume of each of the atmospheric elements in 1 ft³. Obviously, the two largest contributors are $N_2$ and $O_2$ given their proportion by both weight and volume.
- **Line 14** illustrates the total molecular volume for 1 ft³, while **Line 15** demonstrates the volume of the void surrounding all the molecules.
- The math then allows us to calculate the percentage of void, 99.9964%, which means, as an understatement, that there is a whole lot of "nothing" for our quantum of infrared energy of choice to make it through to "infinity and beyond."
- All data shown has properties as noted in the heading because air with moisture is substantially more prevalent than air with no moisture. Note in **Line 12** that the weight of 1 ft³ of dry air weighs more than that same ft³ with 70% moisture saturation. With the exception of Methane ($CH_4$) and Helium (He), water ($H_2O$), is the lightest prevalent molecule in the atmosphere. Therefore, when it is present, it replaces molecules which on the average weigh more, thus the weight of a ft³ is less.

We now need to look at the atmosphere as a ray or photon quantum of infrared energy would see it when it starts its journey from the Earth's surface toward outer space. It goes without saying that the quantum of radiant energy has no idea where it's going to end up. Its total function is to beat-feet the heck-and-gone away from the surface. I employ both the "ray" term and "photon" term because electromagnetic radiation displays characteristics of a ray sometimes and of a photon at other times. In order to be able to put this discussion in perspective, I developed *Figure AP-6: Air at the Beach*. By means of comparison, I have illustrated some of the characteristics of Iron which I will summarize toward the end.

Let's get some dimensions into real-world terms that we can wrap our heads around. I scaled up that "foot" at the bottom of the cubic foot shown, so that it actually can be represented as if it was a line 100 miles long. Now, visualize a beach, 100 miles in length, straight as a surveyor's line, with wave after wave coming in from the ocean. The wave will see "molecules of air" equally spaced at the waterline. They are spaced apart by about 7/8", or a little less than the width of a computer thumb drive. Pretend that they are really straight pins, like a seamstress would employ to sew a garment, stuck upright in the sand, right in line where the wave meets the shore. The smallest molecules, **He** and **Ne**, would be about ½ the diameter of the pin's shank, and the largest molecules, $CH_4$, $CO_2$, $N_2O$ and $H_2O$, would be about the diameter of the head of the pin, ±. Obviously, the wave is constrained very little in its path up the beach. In fact, after that wave passes by the "line of pins," it reforms nicely and you hardly noticed the residue of its encounter with the pins.

Previously, I used the example of the infrared energy being perhaps a boatload of photon canons firing away toward space. Using the above analogy, imagine a 100-foot-wide by 100-foot-high wall with pins positioned as mentioned at 7/8" intervals. Now, stand back about 60 feet and, using your trusty over-and-under, skeet-shooting shotgun loaded with International-Regulation, 12 ga. rounds with #7.5 shot, your task would be to hit one or more of the pinheads with a shotgun blast. I'm not willing to take those odds, because I've missed a bunch of flying, skeet targets that are ~110 mm in diameter, with a face area of 14.75 $in^2$, and they, too, are only 50 to 60 feet away!

A comparison with a 1 $ft^3$ chunk of Iron is appropriate at this point. Iron is the most abundant element on Earth[72], composing 32.1% by mass. The next most abundant element is Oxygen at 30.1%, which by the way, readily combines with Iron to form a "rust," such as $Fe_2O_3$. *Figures AP-6, AP-7 & AP-8* demonstrate some characteristics of Iron as seen in the real world. When Iron changes phase from liquid to solid, it forms crystalline structures as shown in *Figures AP-7 & AP-8*. Polished, raw Iron, has almost no absorptivity or emissivity because of its highly reflective surface. However, the oxidized surface of Iron has a high, 90% ± emissivity; the oxidation process taking place very rapidly once the raw, polished Iron is exposed to atmospheric conditions.

The density of Iron is approximately 485 lbs/$ft^3$, or 222,967 gm/$ft^3$, resulting in 2.405 * $10^{27}$ molecules per $ft^3$ as demonstrated in *Figure AP-6*. Bear in mind that the saturated air in our example contains 34.55 gm/$ft^3$. The center-to-center distance between iron atoms is 2.27 $\text{Å}$, as opposed the molecular-to-molecular distance of 70% saturated air at 34.02 $\text{Å}$. Essentially, those pin-heads mentioned above, when applied to the iron crystals, would be right on top of themselves using the same example. Given the size of all of the molecules involved, Iron has a void density of 18.17% as opposed to the air example of 99.996%. Obviously, this is why we can't "see" through Iron, and most importantly, no Infrared radiation will "pass through" a chunk of iron.

Using another useful comparison, if a wave of infrared energy were to pass through that $ft^3$ of air travelling that 1 foot from bottom to top, it would encounter the 7.19 * $10^{23}$ molecules shown in the figure. Assuming that same ray could penetrate the same $ft^3$ of Iron in same way, it would only have to travel ~0.007" to encounter the same "atomic" resistance. Given that we can't "see through" Iron, my guess is that wave front doesn't even make it that far into the structure.

Lines designated ①, ②, & ③ in *Figure AP-6* show the spacing distances and area percentages of the 7/8" high by 100 mile long "beach" plane. For instance, a $CO_2$ molecule, in 70% saturated air, will occur as 1 out of every 2,439 molecules, with a spacing of 172.5 feet. Water vapor ($H_2O$), on the other hand, occurs in 1 out of every 84 molecules and a spacing of 5.9 feet. Likewise, within that plane, $CO_2$ takes up 0.000104% of the area, and $H_2O$ takes up 0.00239%. By inspection, there is a whole bunch more water vapor in the air than $CO_2$ at these conditions.

# That Shot in the Dark

Now comes the hard part; how to quantify the energy that is radiated back toward the Earth's surface from all those nasty **GHG**s.  Note that I used the term "radiated" as opposed to "reflected."  Reflection occurs with clouds and other "bright" (shiny?) objects, not **GHG** molecules.  Given that clouds are elusive and can never be predicted with any accuracy, the subject of clouds and their effect on the warming of the Earth's surface will not be covered here.  Most importantly, clouds and **GHG**s have little if anything to do with each other, other than their similarity to water vapor.

The science says that it is the **GHG** molecule that absorbs the radiated energy.  This poses two fundamental questions.  The **First Question**, and probably the most important question is how much of that radiant energy does it absorb and what is its resulting "temperature?"  Also, how long does it keep that energy and in what form?  The **Second Question**, which perhaps is equally as important, is how much of that absorbed energy is re-radiated toward, and absorbed by, the surface of the Earth, as opposed to being a major contributing factor to the maintenance of the temperature of our atmosphere?

Okay, let's re-address the S-B equation,

$$P_{emit} = A * \sigma * \varepsilon * T^4$$

and have it deal with two identical, parallel surfaces, $S_1$ & $S_0$, that absorb and emit radiation between themselves. The "$P_{emit}$" equation above only talked about the energy that left the surface, but didn't address anything absorbed by that same surface.  Let's assume that the two surfaces in question have the same area (**A**) and are at the same temperature (**T**), and possess the same emissivity (**ε**). Then by inspection, the energy transferred between the areas in question is identical and no net transfer takes place.

Of great importance is the fact that the surfaces in question are assumed to have infinite thermal masses, infinite ability to transfer heat energy within the mass, and are thermally unaffected by heat loss or gain. In other words, as a function of time, they will continue to transfer energy without changing the emitting/absorbing temperature of either of the surfaces.

Keep in mind going forward that all matter, at a temperature above absolute (-273.15 °C, -459.67 °F), radiates energy per the "$P_{emit}$" equation above, the degree being a function of "**ε**", the emissivity. This function of emissivity assumes both surfaces have the same value, or the areas of one or the other are normalized to the point that the product of (**A** * **ε**) for each surface of the equation is valid.  In other words, when the areas and emissivities are combined, they are normalized to equivalency.

Therefore, back to our two identical surfaces, $S_1$ & $S_0$, if surface $S_1$ is at temperature $T_1$, and surface $S_0$ is at temperature $T_0$, and $T_1$ is greater than $T_0$, then there will be a net exchange of energy from $S_1$ to $S_0$.  Thus, we can quantify the net energy transferred, $P_{net}$, by the following equation.

$P_{net} = P_{emit} - P_{absorb}$

$P_{emit}$ is equal to the value, in Watts or Btu/hour, of the radiant energy emitted outward from $S_1$ and radiated toward $S_0$.

$P_{absorb}$ is the amount of radiant energy $S_1$ absorbs from $S_0$, which of course is the amount of energy emitted by $S_0$.

Once the two equations are combined, the overall equation for the net energy transferred then becomes
$P_{net} = A * \sigma * \varepsilon * (T_1^4 - T_0^4)$

**Where:**

A is the area of the radiating surfaces, assumed to be identical or normalized
$\varepsilon$ is the emissivity/absorptivity of the emitting surfaces, which are assumed to be identical or normalized
$T_1$ is the temperature of the emitting surface, $S_1$, in absolute terms
$T_0$ is the temperature of the body, $S_0$, re-radiating energy back toward $S_1$, again in absolute terms.
$\sigma$ is the Stefan-Boltzmann Constant
$\qquad$ = 5.670374e-08 W/m$^2$/°K$^4$ - SI Dimensions
$\qquad$ = 1.71400e-09 btu/hour/ft$^2$/°R$^4$ - IP Dimensions

Stick with me here, because we're getting to the good part. You will notice that nowhere in this equation, or the S-B law, is distance mentioned; only power, time, absolute temperatures, emissivity and area. Within the plethora of characteristics of the atmosphere that I enumerated for the model described in the book, was the fact that the temperature of the atmosphere decreases as a function of altitude (see *Figure AP-10*). At that point, for simplicity, I told you that at least one of the reasons for the decrease was adiabatic expansion, something I am very familiar with given that I'm an old, refrigeration system design engineer.

Adiabatic expansion, by definition, means that masses of air move upward due to density differentials, expand, and therefore must be replaced from "above." There is no question that some, but by no means all, of the reason for the atmosphere to have any temperature at all, is adiabatic in nature. Also, remember that I told you the differential can't be either conduction or convection from the gases in the atmosphere to any other form of matter beyond the Earth's surface, because the only sink for the energy contained in the air when it leaves the Earth's surface is outer-space, and there literally is nothing there to conduct or convect to.

# The First Question

Now, let's start with the **First Question** having to do with how much energy is absorbed by each molecule of **GHG,** and what temperature does it reach? Inquiring minds wonder if the quantum of energy slated to impact a molecule of **GHG** within, say, the interval of atmosphere between 10,000 feet and 11,000 feet, has a different value than the quantum slated to impact a molecule within, say, the interval between 55,000 feet and 56,000 feet. Per the model, again assuming that the Earth is at a temperature of 59 °F, 15 °C, then the atmospheric temperature at ~10,500 feet is 21.6 °F, or 481.23 °R. Likewise, the atmospheric temperature at ~55,500 feet is -69.7 °F, or 389.97 °R.

Let's further assume, only for clarity purposes (to keep the dimensions within reasonable bounds), that our "molecule" has an area of 1 ft². Employing the S-B equation we can calculate the maximum energy loss from 1 ft² of the Earth's surface to outer space ($T_1$ = 59 °F, 518.67 °R, & $T_0$ = -454.81 °F, 4.86 °R) as being equal to 121.52 btu/hr. Keeping that in mind, then any quantum of energy leaving the surface carries the equivalent heat loss of 121.52 btu/hr., because it has no idea that it is going to "hit" anything but the Moon, best case. Finally, this is not an integral concept; it is a differential; a rate of flow. Effectively, only a "moment in time."

Implicit, hidden in the background of the S-B equation, is the assumption that the two "surfaces," $S_1$ & $S_0$ are three-dimensional, contain an infinite thermal mass, and therefore their temperatures are unaffected by the transfer of radiant energy. In other words, back to that moment in time. However, this is not the case with the "molecule" in our study. It has to get rid of its energy literally as fast as it acquires it; within fractions of a nanosecond probably. [See the **Side-Bar** below] So, if that quantum of energy has to make its way back to the surface of the Earth, thus negating the efforts of the quantum that we thought was classified as "loss," then the S-B equation has to rule, otherwise it, the equation and by inference the law, wasn't valid or a useful tool to begin with.

So, if the steady-state temperature of our 1 ft² molecule at 10,500 ft is 481.23 °R, then S-B says $P_{net}$ = 31.47 btu/hr. If we only lost that amount of energy, and the quantum that provided the heat to begin with had 121.52 btu/hr. potential for loss, then we had to radiate the remainder, $P_{emit}$ = 90.05 btu/hr., back down to Earth from our "molecule." Transposing the S-B equation and solving for $T_1$ we see the following equation that needs resolution.

$$T_1 = [P_{emit} / (A * \varepsilon * \sigma) + T_0{}^4]^{1/4}$$

When we solve for $T_1$ the result turns out to be the fact that instantaneously, our "molecule" had to reach a "temperature" of 136.12 °F, or 595.79 °R in order to "return" radiant energy at that rate. Given what we know about the physical characteristics of the **GHG**s that we are studying, that level of temperature acquisition certainly is not out of the question, at least theoretically on a momentary basis. However, inquiring minds wonder if that really is what happens. There does exist a "hole" in this argument and that has to do with the **Second Question**.

# The Second Question

We should start this portion of the discussion by asking, as a function of time, what was the "temperature" of the **GHG** just prior to being "hit" by the photon/wave of Earth-source radiant energy? Or said a different way, how "hot" does that molecule get before it "unloads" and where does that energy go?

We're going to use a little finite difference logic and approach this problem from the standpoint of "it can't be true." Reference *Table AP-3*. This data centers around a mythical column of 1,000 ft³ of air, with the base sitting on the Earth's surface, and the column being 1,000 feet tall. This example also assumes as a conclusion that all of the radiant energy contained in 1 hour was imparted to the water vapor within the 1,000 feet of column and none of it left the top & escaped to outer space. Remember, the atmosphere is composed of a whole bunch of 1,000-foot columns, so we can assume that nothing "leaves" the sides of the column.

1. **Line 1** gives us the physical constant having to do with the Specific Heat of water vapor, $H_2O$. Note that this is different than liquid water which is ~1.0. **Lines 2 to 6** describe the characteristics of the radiant energy leaving the Earth's surface.
2. **Line 7** further granulates the energy flow from hours to seconds.
3. Knowing the frequency on **Line 6**, we can calculate the btu transmitted per ft² per cycle on **Line 9**.
4. **Line 13** shows us that our 1000 ft³ example of 70% saturated air has 0.55984 lbs. of water vapor contained in it. The derivation of that begins on **Line 10**, *Table AP-2*.
5. The definition of each of the "planes" is that they are composed of packed $H_2O$ molecules side-by-side, 1 molecule thick. Given the number of molecules in 1000 ft³, then the number of planes results in 2,142,841 shown in **Line 15**.
6. **Lines 18** to **24** conclude differentiating the flow of energy such that at the end of the hour, the water vapor in the column, and only the water vapor, reached a temperature of 443.9 °F. Interestingly enough, as shown in **Line 23**, it only takes slightly more than 8 seconds to increase the temperature of all the water vapor 1 °F.

The take-away from this exercise should be obvious: once a **GHG** acquires energy, it has to get rid of it real quick, else it "heats up" disproportionately with respect to its surroundings. Therefore, that energy has to go somewhere, somehow, and most probably as radiation as opposed to conduction or convection.

We talked previously about hemispherical radiation that turned into planes leaving flat surfaces. This is not the case with atmospheric gases. The molecules of **GHG** are functionally spherical in nature and radiate as such. Therefore, almost by definition, whatever energy is acquired by a molecule, only ½ of it heads down & sideways to wherever. The rest travels up and sideways toward outer space and/or destinations unknown.

Let's talk about *Figure AP-9*. Here I've provided three representations of what happens to a methane molecule when it sees radiation emanating from the Earth's surface. As we stated in *Table AP-3*, each "wave" contains the energy equivalent of $1.133 \times 10^{-15}$ btu/cycle/ft$^2$. Above, we talked about the absorbing "surface" having a temperature of 21.6 °F and via the S-B equation, this means that effectively 90.05 btu/hr/ft$^2$ had to be re-radiated by the absorbing unit. Inquiring minds wonder if that molecule of water vapor really was at 21.6 °F when it got "hit?" Remember the basic assumptions when using the S-B regime; the sources and sinks have infinite thermal mass and don't fluctuate in temperature as a function of time. The real questions that we have to address is not whether the molecule gets excited, nor how much energy it releases, it's what happens to that energy, what is the final (rest) state of the emitting molecule and where does the energy end up?

> **Side-Bar Here**. Going forward, you must keep in mind that a Molecule of **GHG** or any of the atmospheric gases for that matter, is not a wind-up toy that "chatters away" for a spell of any length. Unfortunately, we can't observe this process on Tik-Tok or U-Tube, so we have to use a little imagination coupled with hard-core logic. Therefore, its "death rattle" is very short, like nanoseconds, and then it waits for the next "hit." In other words, once the energy is acquired by a molecule, it gets rid of it to its surroundings and in the meantime, picks up more energy via radiation. When the dust settles, you begin to realize that there are a whole bunch of electromagnetic waves wondering around in the atmosphere, containing different energy values, and impacting whatever solid object is available. In fact, all those btu's floating around are fungible. After a while we don't know where they came from nor where they are going. *Figure AP-10* shows you the average results, as determined scientifically, of all those btu's flailing around at various altitudes.

As the elevation increases, the temperature of the atmosphere quickly approaches that which is conducive for radiant energy absorption by $CO_2$. This fact is presented graphically in *Figures B-1, B-2* & *AP-10*. In less than 1,000 feet, it has dropped to a little more than 55 °F, which is well within the range of absorption for $CO_2$. Thus, on the average, we can say that $CO_2$ does not benefit directly from surface temperature radiation on a continuous basis, but must rely upon "seconds" coming from $H_2O$ and "other" molecules. You might ask how much of that second, third or fourth-hand energy reaches the Earth. I suggest that the law of averages indicates very little.

Let's do another thought experiment to help with the meat of this discussion. First of all, re-examining *Table AP-2*, when we compare the total number of **GHG** molecules to the total number of "other" molecules, we see that as a function of mass, the "other" molecules compose 99.19% and **GHG**s compose 0.81% of the total. When viewed from a "number of targets" standpoint, the "other" molecules compose 98.77% and **GHG**s compose 1.23% of the total.

This then can be translated to a conversation about how "target rich" the atmospheric environment really is. By mass, the radiation from any **GHG**, not just $H_2O$, has a 123 out of 124 chances of hitting an "other" molecule, while using the "target" analogy, the ratio is 80 out of 81. Keep in mind our discussion about the S-B equations and the significance of that quantity called emissivity/absorptivity, "**ε,**" a value that can range from 0 to 100%.

Now, the value of **ε** for the **GHG**s in general is between 90% and 98% within the wavelength ranges that I have illustrated. The conclusion to this argument is it only takes an absorptivity of 1% for all the "other" gases to act as receptors and emitters of any and all of that second hand radiation, let alone the "first hand" version of the waveform. The take-away from this discussion is to answer the question of why the temperature of the atmosphere remains as stable as it does, and what is providing the "heat?"

I think it's time we talked about what is depicted in *Figure AP-10*. This is a compilation of data gathered over many years and accepted as "gospel" pertaining to the average of atmospheric conditions and is designated as the U.S. Standard Atmosphere.[73][74] The data used to generate this figure can be found at the footnoted locations. There is much knowledge and fact demonstrated in this simple figure.

1. 50% of the mass of the Earth's atmosphere is below 18,000 ft (5,486 m), which is still within the Troposphere. At that elevation, the temperature is -5.5 °F, and the atmospheric pressure is 14.9 in. Hg, or 7.3 psia.
2. 90% of the mass is below 52,000 ft (15,850 m) which is well into the Tropopause; you remember, that area designated as an inversion layer. The temperature is -69.7 °F, and the atmospheric pressure is 3.95 in. Hg, or 1.92 psia.
3. 99.99997% of the mass is below 328,084 ft (100,000 m), at the Kármán Line, where the "temperature" is -96.07 °F and there is hardly enough pressure to spit at.
4. My models have looked at the atmosphere as high as 234,000 ft (71,000 m), where the "temperature" is -73.3 °F and the atmospheric pressure is 0.00117 in. Hg, or 0.000574 psia. This encompassed 99.9961% of the mass of the Atmosphere.

The 1995-1996 edition of the Handbook of Chemistry and Physics that I have referenced has 5 pages, 14-18 to 14-23, devoted to nothing but graphs depicting atmospheric molecular activity. The terminology used is very scientific and beyond the scope of this book. However, the takeaway is simple to express. Whatever radiant energy enters the atmosphere, stays in the atmosphere, or is expelled into the great beyond, and the odds of it returning to Earth are slim to nil.

The effect of this expelled, Earth-surface generated energy on the increase in that surface temperature is classically "in the noise," as shown in *Figures B-7 & B-8*. I stated then that I was very generous, and conservative, when modelling the **GHG** effect, and in fact gave it far more credit than it was due. Therefore, given the above argument, we can only conclude that it is the heat generated through the combustion of fossil fuels or biomass, or the products of fission in nuclear reactors, that incrementally add heat the Earth.

# ... and in Summary,

To put a fine point on this issue, we have been Zohnerized [see **Have we been "42'd" (\*)?**] by the U.N. and the IPCC. They have pursued an agenda based upon correlation as opposed to causality. We are not questioning the fact that the Earth is warming. Nor should we question that the atmospheric content of $CO_2$ is increasing. The increase in $CO_2$ is a byproduct, not a cause. It's not the "ifs" that are important, it's the "whys." But most importantly, when all the dust settles, it will turn out that the increase in $CO_2$ will be a "so what" or "who cares" issue. In fact, we are going to want that additional $CO_2$ in the coming years.

The Green New Dealers are promoting a crusade that really has no substance. Perhaps, we should put our "thinking caps" on, trace the money, and then understand who benefits from all the hoopla. I can almost guarantee that we, the average tax-paying citizens of the United States, are putting money from our pockets into someone else's pocket and getting nothing but grief in return. Do mask & VAX mandates ring a bell?

As I have emphatically stated before, $CO_2$ is our friend, not our enemy. It is the life-blood of our human existence, flowing through our lives like a "C" of Carbon. Therefore, leave it alone and get on with the solution to the fundamental problem that all of humanity faces; we are running out of fossil fuels. Let's use our heads and come up with rational solutions to that problem and quit creating strawman problems for someone else's aggrandizement. And, OBTW, in the process, get all the government entities out of the way.

Q.E.D., *quod erat demonstrandum*, Latin: **which was to be demonstrated**.

# Tables and Figures Appendix A

## Figure A-1
### U.S. Energy Consumption by Source and Sector, 2020

Sources: U.S. Energy Information Administration
*Monthly Energy Report* (April 2021, Table 1.3)

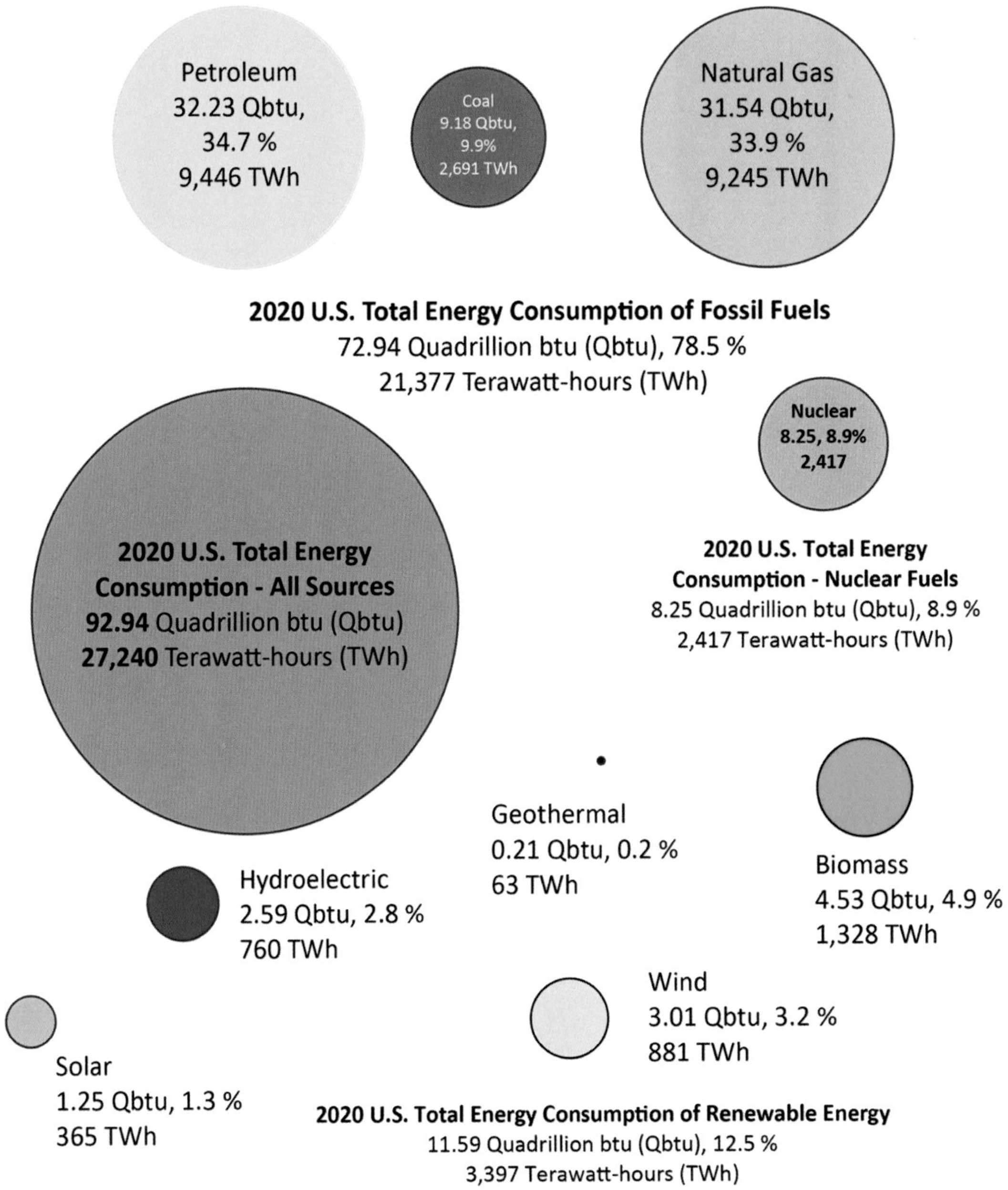

Petroleum
32.23 Qbtu,
34.7 %
9,446 TWh

Coal
9.18 Qbtu,
9.9%
2,691 TWh

Natural Gas
31.54 Qbtu,
33.9 %
9,245 TWh

**2020 U.S. Total Energy Consumption of Fossil Fuels**
72.94 Quadrillion btu (Qbtu), 78.5 %
21,377 Terawatt-hours (TWh)

Nuclear
8.25, 8.9%
2,417

**2020 U.S. Total Energy**
**Consumption - All Sources**
**92.94** Quadrillion btu (Qbtu)
**27,240** Terawatt-hours (TWh)

**2020 U.S. Total Energy**
**Consumption - Nuclear Fuels**
8.25 Quadrillion btu (Qbtu), 8.9 %
2,417 Terawatt-hours (TWh)

Geothermal
0.21 Qbtu, 0.2 %
63 TWh

Biomass
4.53 Qbtu, 4.9 %
1,328 TWh

Hydroelectric
2.59 Qbtu, 2.8 %
760 TWh

Wind
3.01 Qbtu, 3.2 %
881 TWh

Solar
1.25 Qbtu, 1.3 %
365 TWh

**2020 U.S. Total Energy Consumption of Renewable Energy**
11.59 Quadrillion btu (Qbtu), 12.5 %
3,397 Terawatt-hours (TWh)

## Table AP-1, Atmospheric Particle Absorption of Infrared Radiation
# Infrared Absorption Characteristics of Earth's Atmosphere
### CRC, Handbook of Chemistry and Physics, 1995 - 1996 edition
#### pp 14-22 & 10-293

| °C | °K | °F | °R | meV | THz | µm |
|------|-------|-----|-------|-------|------|-------|
| -17.8 | 255.4 | 0 | 459.7 | 109.5 | 26.4 | 11.35 |
| -15.6 | 257.6 | 4 | 463.7 | 110.5 | 26.7 | 11.25 |
| -13.3 | 259.8 | 8 | 467.7 | 111.4 | 26.9 | 11.15 |
| -11.1 | 262.0 | 12 | 471.7 | 112.3 | 27.1 | 11.06 |
| -8.9 | 264.3 | 16 | 475.7 | 113.3 | 27.3 | 10.96 |
| -6.7 | 266.5 | 20 | 479.7 | 114.2 | 27.6 | 10.87 |
| -4.4 | 268.7 | 24 | 483.7 | 115.2 | 27.8 | 10.78 |
| -2.2 | 270.9 | 28 | 487.7 | 116.1 | 28.0 | 10.70 |
| 0.0 | 273.2 | 32 | 491.7 | 117.1 | 28.3 | 10.61 |
| 2.2 | 275.4 | 36 | 495.7 | 118.0 | 28.5 | 10.52 |
| 4.4 | 277.6 | 40 | 499.7 | 119.0 | 28.7 | 10.44 |
| 6.7 | 279.8 | 44 | 503.7 | 119.9 | 28.9 | 10.36 |
| 8.9 | 282.0 | 48 | 507.7 | 120.9 | 29.2 | 10.27 |
| 11.1 | 284.3 | 52 | 511.7 | 121.8 | 29.4 | 10.19 |
| 13.3 | 286.5 | 56 | 515.7 | 122.8 | 29.6 | 10.12 |
| 15.6 | 288.7 | 60 | 519.7 | 123.7 | 29.9 | 10.04 |
| 17.8 | 290.9 | 64 | 523.7 | 124.7 | 30.1 | 9.96 |
| 20.0 | 293.2 | 68 | 527.7 | 125.6 | 30.3 | 9.89 |
| 22.2 | 295.4 | 72 | 531.7 | 126.6 | 30.6 | 9.81 |
| 24.4 | 297.6 | 76 | 535.7 | 127.5 | 30.8 | 9.74 |
| 26.7 | 299.8 | 80 | 539.7 | 128.4 | 31.0 | 9.67 |
| 28.9 | 302.0 | 84 | 543.7 | 129.4 | 31.2 | 9.60 |
| 31.1 | 304.3 | 88 | 547.7 | 130.3 | 31.5 | 9.53 |
| 33.3 | 306.5 | 92 | 551.7 | 131.3 | 31.7 | 9.46 |
| 35.6 | 308.7 | 96 | 555.7 | 132.2 | 31.9 | 9.39 |
| 37.8 | 310.9 | 100 | 559.7 | 133.2 | 32.2 | 9.32 |
| 40.0 | 313.2 | 104 | 563.7 | 134.1 | 32.4 | 9.26 |
| 42.2 | 315.4 | 108 | 567.7 | 135.1 | 32.6 | 9.19 |
| 44.4 | 317.6 | 112 | 571.7 | 136.0 | 32.8 | 9.13 |
| 46.7 | 319.8 | 116 | 575.7 | 137.0 | 33.1 | 9.06 |
| 48.9 | 322.0 | 120 | 579.7 | 137.9 | 33.3 | 9.00 |
| 51.1 | 324.3 | 124 | 583.7 | 138.9 | 33.5 | 8.94 |
| 53.3 | 326.5 | 128 | 587.7 | 139.8 | 33.8 | 8.88 |
| 55.6 | 328.7 | 132 | 591.7 | 140.8 | 34.0 | 8.82 |
| 57.8 | 330.9 | 136 | 595.7 | 141.7 | 34.2 | 8.76 |

Water Vapor, H₂O, 4.9µm to 13µm - 2.2µm to 3.45µm

Carbon Dioxide, CO₂ 10.1µm to 13+µm, 4.9µm to 4.05µm

CO₂ 9.1µm to 9.6µm minimal Significance

N₂O & CH₄

| | Low Temperature Boundary | | | High Temperature Boundary | | |
|------|------|------|-----|------|------|-----|
| | °F | THz | µm | °F | THz | µm |
| H₂O | -58.6 | 23.1 | >13 | 604 | 61.2 | 4.9 |
| CO₂ | -58.6 | 23.1 | >13 | 56.9 | 29.7 | 10.1 |
| N₂O | 133.1 | 34.1 | 8.8 | 231.5 | 39.7 | 7.55 |
| CH₄ | 150.6 | 35.1 | 8.55 | 408.2 | 42.2 | 7.1 |

## Figure AP-1: Water Molecule

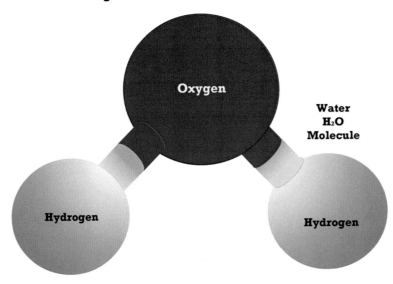

## Figure AP-2: Carbon Dioxide Molecule

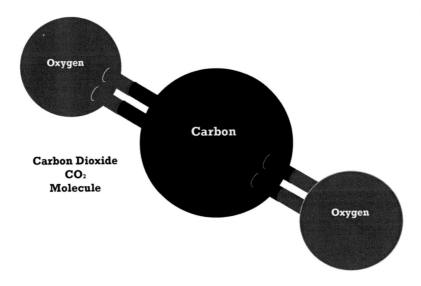

## Figure AP-3: Methane Molecule

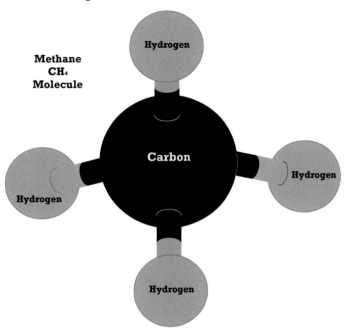

Methane
CH₄
Molecule

## Figure AP-4: Nitrous Oxide Molecule

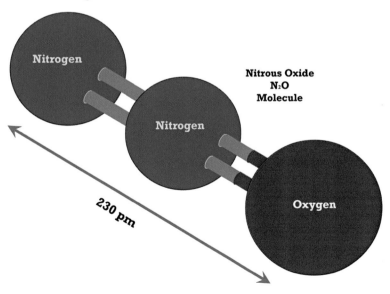

Nitrous Oxide
N₂O
Molecule

## Figure AP-5: Oxygen Molecule

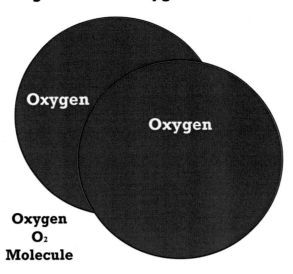

Oxygen
$O_2$
Molecule

## Table AP-2: Atmospheric Molecular Statistics

| | | | | Molecules | | | | |
|---|---|---|---|---|---|---|---|---|
| Name | Formula | gram per ft³ | mol per ft³ | per ft³ | diameter Å | Volume Å³ | % vol - area | line |
| Nitrogen | $N_2$ | 25.92170858 | 0.92116946 | 5.54741E+23 | 1.42 | 8.31676E+23 | 0.002937036% | 1 |
| Oxygen | $O_2$ | 7.90681197 | 0.24710332 | 1.48809E+23 | 1.26 | 1.55862E+23 | 0.000550421% | 2 |
| Argon | Ar | 0.44019098 | 0.01101855 | 6.63552E+21 | 1.06 | 4.13801E+21 | 0.000014613% | 3 |
| Carbon Dioxide | $CO_2$ | 0.02154602 | 0.00048958 | 2.94833E+20 | 1.93 | 1.10981E+21 | 0.000003919% | 4 |
| Neon | Ne | 0.00043281 | 0.00002145 | 1.29158E+19 | 0.58 | 1.31949E+18 | 0.000000005% | 5 |
| Helium | He | 0.00002474 | 0.00000618 | 3.72271E+18 | 0.28 | 4.2789E+16 | 0.000000000% | 6 |
| Methane | $CH_4$ | 0.00003539 | 0.00000221 | 1.32853E+18 | 1.31 | 1.5638E+18 | 0.000000006% | 7 |
| Krypton | Kr | 0.00011270 | 0.00000134 | 8.09903E+17 | 1.16 | 6.6192E+17 | 0.000000002% | 8 |
| Nitrous Oxide | $N_2O$ | 0.00001734 | 0.00000039 | 2.36577E+17 | 2.31 | 1.53086E+18 | 0.000000005% | 9 |
| Water Vapor | $H_2O$ | 0.25664667 | 0.01424628 | 8.57931E+21 | 1.72 | 2.28122E+22 | 0.000080561% | 10 |
| **Total 70% Saturated** | | 34.54753 | | 7.19079E+23 | | 1.016E+24 | | 11 |
| Total Dry Air | | 34.70223 | | | | | | 12 |
| Total Volume, 1 ft³ | | 2.832E+28 | Å³ | | Total GHG = 0.0000844908% | | | 13 |
| Molecular Volume | | 1.016E+24 | Å³ | | | | | 14 |
| Void volume, Å³ | | 2.832E+28 | Å³ | | | | | 15 |
| % Void | | 99.9964% | | | | | | 16 |

Atmospheric Composition, 70% saturated air, 59 °F, 14.696 psia
Angstroms per foot = 3,048,000,000

## Figure AP-6: Air at the Beach

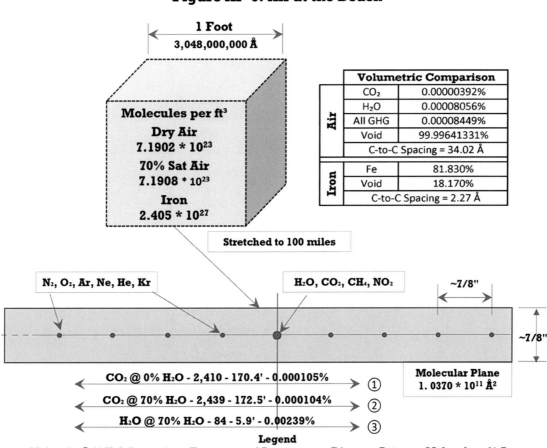

**1 Foot**
3,048,000,000 Å

**Molecules per ft³**
**Dry Air**
$7.1902 * 10^{23}$
**70% Sat Air**
$7.1908 * 10^{23}$
**Iron**
$2.405 * 10^{27}$

| Volumetric Comparison | | |
|---|---|---|
| Air | $CO_2$ | 0.00000392% |
| | $H_2O$ | 0.00008056% |
| | All GHG | 0.00008449% |
| | Void | 99.99641331% |
| | C-to-C Spacing = 34.02 Å | |
| Iron | Fe | 81.830% |
| | Void | 18.170% |
| | C-to-C Spacing = 2.27 Å | |

**Stretched to 100 miles**

$N_2$, $O_2$, Ar, Ne, He, Kr

$H_2O$, $CO_2$, $CH_4$, $NO_2$

~7/8"

~7/8"

**Molecular Plane**
$1.0370 * 10^{11}$ Å²

CO₂ @ 0% H₂O - 2,410 - 170.4' - 0.000105%  ①
CO₂ @ 70% H₂O - 2,439 - 172.5' - 0.000104%  ②
H₂O @ 70% H₂O - 84 - 5.9' - 0.00239%  ③

**Legend**
**Molecule @ % H₂O Saturation - Frequency of Appearance - Distance Between Molecules - % Area**

**Figure AP-7: Face-Centered Iron Crystal**

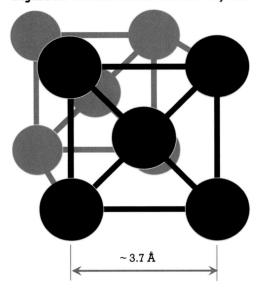

~ 3.7 Å

**Figure AP-8: Body-Centered Iron Crystal**

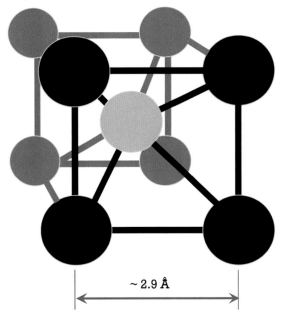

~ 2.9 Å

# Table AP-3
# Radiant Energy Impact, 1st 1000 feet

| | | |
|---|---:|---:|
| Specific Heat of $H_2O$ vapor, btu/lb./°F | 0.489 | 1 |
| Surface Temperature of the Earth | 59 °F | 2 |
| Temperature of Outer Space | 4.86 °R | 3 |
| Radiation from 59 °F surface, btu/hr/ft² | 121.52 | 4 |
| Wavelength, µm | 10.06 | 5 |
| Frequency, THZ | 29.8 | 6 |
| btu/second/ft² | 3.3756E-02 | 7 |
| seconds per cycle | 3.3557E-14 | 8 |
| btu/cycle/ft² | 1.13274E-15 | 9 |
| Area Å² per Molecule, *Table AP-2* | 2.3204 | 10 |
| Å² per ft² | 9.2903E+18 | 11 |
| molecules per ft² plane | 4.00371E+18 | 12 |
| lbs $H_2O$ vapor per 1000 ft³ | 0.55984 | 13 |
| Molecules $H_2O$ per 1000 ft³, *Table AP-2* | 8.57931E+24 | 14 |
| Number of Planes per 1000 ft | 2,142,841 | 15 |
| Gap Between Planes, Å | 1,437,567 | 16 |
| Gap Between Planes, inches | 0.00566 | 17 |
| lbs $H_2O$ per Plane | 2.613E-07 | 18 |
| btu/ft² plane/°F | 1.29118E-07 | 19 |
| °F/cycle/plane | 8.77285E-09 | 20 |
| cycles/°F/plane | 113,988,022 | 21 |
| Seconds/°F/plane | 3.8251E-06 | 22 |
| Seconds/°F/ for all planes | 8.11 | 23 |
| °F/hour/1000 ft³ | 443.9 | 24 |

# Figure AP-9: Shake, Rattle & Radiate

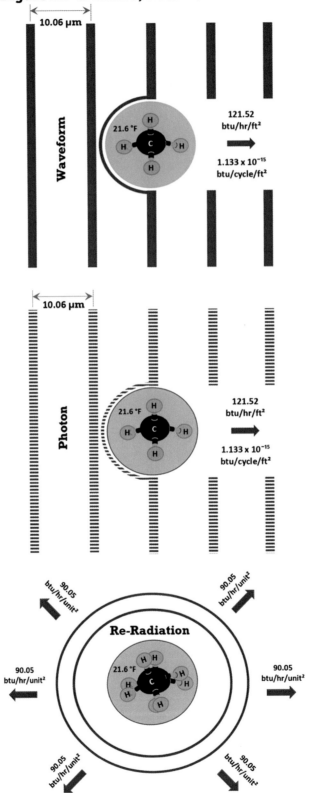

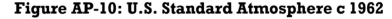

# Figure AP-10: U.S. Standard Atmosphere c 1962

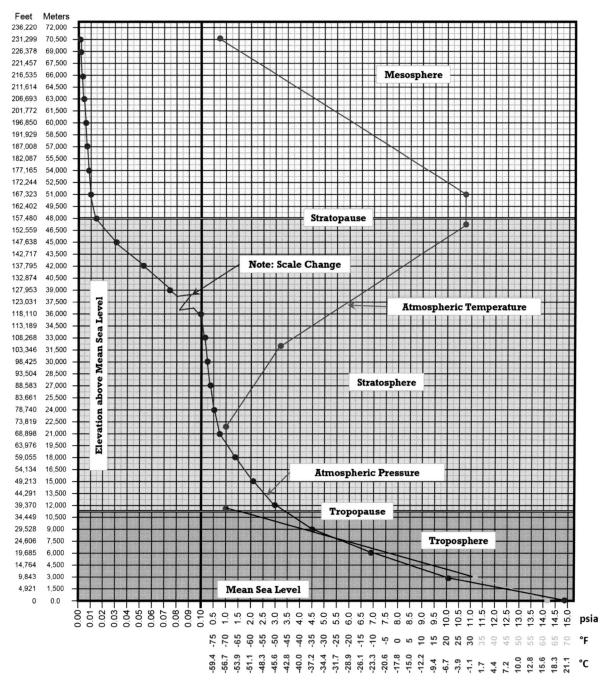

**Table AP-4: Atmospheric Electromagnetic Energy Comparisons**

| nm | μm | °K | °C | °R | °F | THz | watts/m² | Total | per Wave | power | # "hits" | |
|---|---|---|---|---|---|---|---|---|---|---|---|---|
| 380 | 0.38 | 7,625 | 7352 | 13725 | 13265 | 788.93 | 1.9E+08 | 6.1E+07 | 2.1E-11 | x 18521 | 26.5 | Visible |
| 700 | 0.70 | 4,141 | 3868 | 7454 | 6994 | 428.27 | 1.7E+07 | 5.3E+06 | 3.4E-12 | x 2968 | 14.4 | Light |
| 900 | 0.90 | 3,221 | 2948 | 5798 | 5338 | 333.10 | 6.1E+06 | 1.9E+06 | 1.6E-12 | x 1397 | 11.2 | |
| 1100 | 1.10 | 2,634 | 2361 | 4741 | 4282 | 272.54 | 2.7E+06 | 8.7E+05 | 8.8E-13 | x 763 | 9.1 | Infrared |
| 1300 | 1.30 | 2,229 | 1956 | 4012 | 3553 | 230.61 | 1.4E+06 | 4.4E+05 | 5.3E-13 | x 463 | 7.7 | "A" |
| 1500 | 1.50 | 1,932 | 1659 | 3478 | 3018 | 199.86 | 7.9E+05 | 2.5E+05 | 3.5E-13 | x 301 | 6.7 | |
| 8000 | 8.00 | 362 | 89 | 652 | 192 | 37.47 | 9.7E+02 | 3.1E+02 | 2.3E-15 | x 2.0 | 1.3 | |
| 9000 | 9.00 | 322 | 49 | 580 | 120 | 33.31 | 6.1E+02 | 1.9E+02 | 1.6E-15 | x 1.4 | 1.1 | Infrared |
| 10000 | 10.00 | 290 | 17 | 522 | 62 | 29.98 | 4.0E+02 | 1.3E+02 | 1.2E-15 | x 1.0 | 1.0 | "C" |
| 10060 | 10.06 | 288.2 | 15 | 519 | 59 | 29.80 | 3.9E+02 | 1.2E+02 | 1.2E-15 | x 1 | 1 | |
| 11000 | 11.00 | 264 | -9 | 475 | 15 | 27.25 | 2.8E+02 | 8.7E+01 | 8.9E-16 | x 0.8 | 0.9 | |
| 1 | 2 | 3 | 4 | 5 | 6 | 7 | 8 | 9 | 10 | 11 | 12 | Column |

Header groupings:

| Wave Length | | Temperatures | | | | Freq. | Power — Total | Power — btu/hr/ft² | | Power & Frequency Normal to 59 °F | | |
|---|---|---|---|---|---|---|---|---|---|---|---|---|
| nm | μm | °K | °C | °R | °F | THz | watts/m² | Total | per Wave | power | # "hits" | |

# Glossary of Terms:

| | |
|---|---|
| Billion | $10^9$ of anything |
| Million | $10^6$ of anything |
| Quadrillion | $10^{15}$ of anything |
| Trillion | $10^{12}$ of anything |
| ASHRAE | American Society of Heating, Refrigerating and Air Conditioning Engineers |
| BLM | Black Lives Matter |
| BYA | Billion Years Ago |
| CERES | Clouds and the Earth's Radiant Energy System |
| CHP | Combined Heat and Power |
| EIA | U.S. Energy Information Administration |
| EPA | U.S. Environmental Protection Agency |
| EV | Electric Vehicle; various % of Electric Power usage |
| GE, **GE** | Greenhouse Effect |
| GHG | Greenhouse Gas |
| GND, **GND** | Green New Deal |
| HF | Human Factor |
| IP | Inch-pound System of Units |
| IPCC | Intergovernmental Panel on Climate Change |
| MPG | Miles per Gallon |
| MYA | Million Years Ago |
| NASA | National Aeronautics and Space Administration |
| NIMB | Not in my Backyard |
| NOAA | National Oceanic and Atmospheric Administration |
| OBTW | Oh, by the way |
| OPEC | Organization of Petroleum Exporting Countries |
| RF | Radiative Forcing |
| S-B | Stefan-Boltzmann Constant/Law |
| SI | International System of Units |
| Speed of Light, **c** | 186,282 miles per second; 299,792,458 meters per second |
| TOU | Time-of-Use, primarily as applied to Electric Power Usage |
| TPES | Total Primary Energy Supply |
| WRT | With respect to |

| | |
|---|---|
| °C | degree Celsius |
| °F | degree Fahrenheit |
| °K | degree Kelvin |
| °R | degree Rankine |
| μm | $= 10^{-6}$ meters, $10^4$ Å, Angstroms |
| Å | Angstrom $= 10^{-10}$ meters |
| Angstrom | $= 10^{-10}$ meters, Å |
| Avogadro Number | $6.02214076 * 10^{23}$ |
| Btu, btu | British Thermal Unit: the energy to raise 1 lb. of water 1 °F |
| $CH_4$ | Methane |
| $CO_2$ | Carbon Dioxide |
| eV | electron volt $= 1.6022 * 10^{-19}$ J |
| fm | femtometer, $= 10^{-15}$ meters |
| ft | foot, 0.3048 meter |
| $ft^2$ | square foot |
| $ft^3$ | cubic foot |
| $H_2$ | Hydrogen molecule |
| $H_2O$ | Dihydrogen Monoxide, water |
| ha | hectare, 10,000 $m^2$, 107,636.5 $ft^2$ |
| Hg | Inches of Mercury |
| hr | hour |
| J | Joule, $= 1$ Ws; 3600 J/Wh |
| Joule | $= 1$ Ws |
| kJ | $= 1,000$ Joules |
| kW | Kilowatt: 1,000 Watts, 3,412 btu/hr |
| kWh | Kilowatt Hour: 3,412.14 btu |
| lb. | pound weight |
| m | meter, 3.2808 ft |
| $m^2$ | square meter |
| $m^3$ | cubic meter |
| meV | Million Electron Volts |
| microsecond | $= 10^{-6}$ seconds, μs |
| $mile^2$ | Square mile, 27,878,400 $ft^2$ |
| mol | a quantity of molecules or atoms equaling Avogadro's number |
| $N_2$ | Nitrogen molecule |
| $N_2O$ | Nitrous Oxide |
| nanosecond | $= 10^{-9}$ seconds |
| $O_2$ | Oxygen molecule |
| $O_3$ | Ozone |

| | |
|---|---|
| picosecond | $= 10^{-12}$ seconds |
| pm | picometer, $10^{-12}$ meters |
| ppmm | parts per million by mass |
| ppmv | parts per million by volume |
| psia | pounds per square inch, absolute |
| psig | pounds per square inch, gauge |
| Qbtu | Quadrillion btu, $10^{15}$ btu |
| TeV | Trillion Electron Volts, $1.6022 * 10^{-7}$ J |
| TWh | Terawatthour, 293.08 Qbtu, 1 billion kWhs |
| W | Watt: 0.293071 btu/hr |
| Wh | Watthour; 3,600 Joules |
| Ws | Watt Second |
| ε | Emissivity/absorptivity symbol |
| σ | Stefan-Boltzmann Constant symbol |

# (Endnotes)

1. https://en.wikipedia.org/wiki/Plastic
2. https://en.wikipedia.org/wiki/Airborne_fraction
3. https://en.wikipedia.org/wiki/Greenhouse_gas
4. https://en.wikipedia.org/wiki/Black-body_radiation
5. https://en.wikipedia.org/wiki/Greenhouse_gas
6. https://en.wikipedia.org/wiki/Atmosphere_of_Earth
7. https://en.wikipedia.org/wiki/Cloud
8. https://en.wikipedia.org/wiki/Plastic
9. https://en.wikipedia.org/wiki/Photosynthesis
10. https://www.ncbi.nlm.nih.gov/pmc/articles/PMC2195461/
11. https://www.merriam-webster.com/dictionary/weather
12. https://www.merriam-webster.com/dictionary/climate
13. https://en.wikipedia.org/wiki/Carbon
14. https://en.wikipedia.org/wiki/Greenhouse_gas
15. https://en.wikipedia.org/wiki/Carbon_dioxide
16. https://en.wikipedia.org/wiki/Greenhouse_gas
17. https://www.theverge.com/2021/6/7/22522736/carbon-dioxide-co2-record-climate-change
18. https://en.wikipedia.org/wiki/Carbon_dioxide
19. https://en.wikipedia.org/wiki/Greenhouse_gas
20. https://en.wikipedia.org/wiki/Electricity_generation#Generators
21. https://en.wikipedia.org/wiki/Electricity_generation#Generators
22. https://en.wikipedia.org/wiki/Uranium
23. https://en.wikipedia.org/wiki/Uranium
24. https://en.wikipedia.org/wiki/Wind_power
25. https://en.wikipedia.org/wiki/Geothermal_gradient
26. https://en.wikipedia.org/wiki/Primary_energy
27. https://en.wikipedia.org/wiki/Primary_energy
28. https://en.wikipedia.org/wiki/Solar_panel
29. https://en.wikipedia.org/wiki/Wind_power
30. https://en.wikipedia.org/wiki/Clouds_and_the_Earth%27s_Radiant_Energy_System
31. https://en.wikipedia.org/wiki/Climate_change
32. https://www.merriam-webster.com/dictionary/weather
33. https://www.merriam-webster.com/dictionary/climate
34. https://en.wikipedia.org/wiki/Solar_constant
35. https://en.wikipedia.org/wiki/Albedo
36. https://en.wikipedia.org/wiki/Infrared
37. *https://en.wikipedia.org/wiki/Solar_irradiance/Irradiance on Earth's surface*
38. NextEra Energy, Inc. Environmental, Social and Governance Report 2021, pg 67
39. https://en.wikipedia.org/wiki/Solar_panel
40. https://www.census.gov/geographies/reference-files/2010/geo/state-area.html
41. https://en.wikipedia.org/wiki/World_population
42. *https://en.wikipedia.org/wiki/Demographics_of_the_United_States*
43. https://en.wikipedia.org/wiki/Clouds_and_the_Earth%27s_Radiant_Energy_System

44. https://en.wikipedia.org/wiki/Troposphere
45. https://en.wikipedia.org/wiki/Atmospheric_temperature
46. https://en.wikipedia.org/wiki/Troposphere
47. https://en.wikipedia.org/wiki/Black-body_radiation
48. https://en.wikipedia.org/wiki/Stefan%E2%80%93Boltzmann_law
49. https://en.wikipedia.org/wiki/Black-body_radiation
50. https://en.wikipedia.org/wiki/Radiative_forcing
51. *Marks' Mechanical Engineering Handbook, 6th Edition, pp 4-108 through 4-121.*
52. https://en.wikipedia.org/wiki/Greenhouse_and_icehouse_Earth#Greenhouse_Earth
53. https://en.wikipedia.org/wiki/Great_Oxidation_Event
54. *Study pinpoints timing of oxygen's first appearance in Earth's atmosphere _ MIT News _ Massachusetts Institute of Technology*
55. https://en.wikipedia.org/wiki/Geothermal_gradient
56. *https://en.wikipedia.org/wiki/Earth*
57. https://web.archive.org/web/20210512202029/https://archive.ipcc.ch/publications_and_data/publications_and_data_reports.shtml
58. http://agage.mit.edu
59. *https://www.farmprogress.com/florida-farmland-values-soaring-0*
60. *https://en.wikipedia.org/wiki/Brownian_motion*
61. https://en.wikipedia.org/wiki/Atmosphere_of_Earth
62. *https://en.wikipedia.org/wiki/Infrared*
63. *CRC, Handbook of Chemistry and Physics, 1995 - 1996 edition, pg 10-293, Black Body Radiation*
64. *CRC, Handbook of Chemistry and Physics, 1995 - 1996 edition, pg 14-22*
65. *https://en.wikipedia.org/wiki/Albedo*
66. *https://en.wikipedia.org/wiki/Electronvolt*
67. *https://en.wikipedia.org/wiki/Electronvolt*
68. https://en.wikipedia.org/wiki/Black-body_radiation
69. https://en.wikipedia.org/wiki/Stefan%E2%80%93Boltzmann_law
70. *https://en.wikipedia.org/wiki/Speed_of_light*
71. *https://en.wikipedia.org/wiki/Atmosphere_of_Earth*
72. *https://en.wikipedia.org/wiki/Iron*
73. *https://en.wikipedia.org/wiki/Atmosphere_of_Earth*
74. https://www.centennialofflight.net/essay/Theories_of_Flight/atmosphere/TH1G1.htm

Printed in the United States
by Baker & Taylor Publisher Services